Edwin James Houston, Arthur Edwin Kennelly

Electric Incandescent Lighting

Edwin James Houston, Arthur Edwin Kennelly

Electric Incandescent Lighting

ISBN/EAN: 9783337249328

Printed in Europe, USA, Canada, Australia, Japan

Cover: Foto ©berggeist007 / pixelio.de

More available books at **www.hansebooks.com**

ELEMENTARY ELECTRO-TECHNICAL SERIES

ELECTRIC INCANDESCENT LIGHTING

BY

EDWIN J. HOUSTON, Ph. D.

AND

A. E. KENNELLY, Sc. D.

NEW YORK
THE W. J. JOHNSTON COMPANY
253 BROADWAY
1896

PREFACE.

This little volume has been prepared by the authors with the view of presenting both the art and science of practical electric incandescent lighting to the general public, in such a manner as shall render it capable of being understood without any previous technical training.

The necessity of some general knowledge of the principles underlying the practical applications of incandescent lighting, will be appreciated from the fact that at the present time incandescent lamps are manufactured in the United States at a rate of about eight millions per annum. This of course represents a large amount of invested capital, only a small portion of which is engaged in the actual manufacture of the lamps, by far the greater part being invested in the central stations where the electric current is generated, and in the streets and buildings where the conductors and fixtures are provided. Since the whole of this important industry has practically come into existence since 1881, comparatively little opportunity has been afforded to the public of acquiring a fair understanding of the subject.

It is in the hope of meeting the above want that the authors have written this book.

PHILADELPHIA,
June 1, 1896.

CONTENTS.

CHAPTER		PAGE
I.	Artificial Illumination,	1
II.	Early History of Incandescent Lighting,	18
III.	Elementary Electrical Principles,	43
IV.	Physics of the Incandescent Electric Lamp,	65
V.	Manufacture of Incandescent Lamps. Preparation and Carbonization of the Filament,	83
VI.	Mounting and Treatment of Filaments,	99
VII.	Sealing-in and Exhaustion,	118
VIII.	Lamp Fittings,	135
IX.	The Incandescent Lamp,	163
X.	Light and Illumination,	198
XI.	Systems of Lamp Distribution,	210
XII.	House Fixtures and Wiring,	237

CHAPTER		PAGE
XIII.	Street Mains,	284
XIV.	Central Stations,	304
XV.	Isolated Plants,	324
XVI.	Meters,	334
XVII.	Storage Batteries,	345
XVIII.	Series Incandescent Lighting,	374
XIX.	Alternating-Current Circuit Incandescent Lighting,	386
XX.	Miscellaneous Applications of Incandescent Lamps,	402
Index,		419

ELECTRIC INCANDESCENT LIGHTING.

CHAPTER I.

ARTIFICIAL ILLUMINATION.

Doubtless, the earliest artificial illuminant employed by primitive man was the blazing fagot, seized from the fire. The step from this to the oil lamp marked an era in civilization, since man's work was then not necessarily limited in time by the rising and setting of the sun. It is difficult, at this time, to estimate properly the great boon to civilization this invention afforded. Although at first sight it does not seem to be a very great step from the flickering light of the torch to that of the oil lamp, yet, when we consider the requirements of an artificial illuminant, the superiority of the latter is evident in what may, perhaps, be regarded as the most essential requirement of such an illuminant; namely, its duration; *i. e.*, the length of time during which it can supply a proper illumination without renewal. Instead of the fitful flickering of the torch we have the comparatively steady glow of the oil lamp; instead of the evanescent light of the torch, we have in the oil lamp a means for furnishing light for many hours.

Fortunately, for the sake of progress, man's ingenuity was not arrested by this great achievement. There followed in its wake, various improvements in forms of oil lamps, but this illuminant sufficed to light the world for many centuries, as the tombs, parchments and bas reliefs of the remote past attest.

Passing by the many improvements in various forms of oil lamps, perhaps the next noted step in the production of artificial light followed in the discovery of coal oil, and marked improvements were made in lamps designed especially to burn this natural illuminant. The next great improvement in this direction was the lighting of extended areas by means of illuminating gas. Here, for the first time in the history of the art, means were devised for supplying an illuminant from a central station, under circumstances which permitted of its ready distribution over extended areas.

The greatest step, however, in the production of an artificial illuminant was undoubtedly that which followed the invention of the voltaic pile in 1796. Then, for the first time in the history of science, means were provided whereby powerful electric currents could be readily produced and their effects observed. Naturally the use of these currents soon led to the discovery of the very powerful illuminating properties possessed by the voltaic arc. It must not be supposed, however, that in these successive steps each new illuminant completely supplanted its predecessors. On the contrary, the very fact that night could thus be transformed into day, stimulated improvements in the pre-existing forms of illuminants, and, in the emulation thus produced, marked advances were made in earlier methods. Thus, it was at one time claimed, when gas was first introduced into cities, that existing methods of

ARTIFICIAL ILLUMINATION.

lighting would soon be entirely replaced by the new illuminant, but, so far from this being the case, old methods were sufficiently improved to fairly hold their own, and even to require continued improvements in the new illuminant, in order to maintain its superiority. So, again, when the electric light was introduced, it was predicted by enthusiasts that gas lighting would now be entirely replaced by its new rival, but, as is well known, this expectation was not realized. So markedly have improvements in different methods of illumination gone hand in hand, that we possess to-day all our original methods, save, perhaps, the torch.

Before discussing the advantages possessed by the incandescent electric light, it will be advantageous to consider in general the requirements of any artificial illuminant. It is evident that the most satisfactory artificial illuminant will be that whose properties most nearly resemble those of the sunlight it replaces. No existing illuminant completely satisfies our requirements from this standpoint. The ideal artificial illuminant should possess the following properties; it should be,

(1) Safe.
(2) Cheap.
(3) Hygienic.
(4) Steady.
(5) Reliable.
(6) Akin to sunlight in color.
(7) Capable of ready subdivision.
(8) Cool.
(9) Readily turned on and off at a distance.
(10) Amenable to the purposes of decoration.

As regards safety, it is evident that an artificial illuminant should be safe both to property and to life. All bare or naked flames are open to the objection that combustible material, coming in contact with them, may start dangerous conflagrations.

A good artificial illuminant must necessarily be cheap; for, unless it can be furnished at a price which brings it fairly within the reach of all, its use will be very limited.

All artificial illuminants must necessarily permit of continued use without any deleterious effects on the health. Such defects may arise either from products of combustion vitiating the surrounding air, or, possibly, in extreme cases, from injury to sight.

Steadiness is a prime essential of a good

illuminant. If an artificial light produces an unsteady, flickering illumination, an injurious strain may be put on the eye, in its endeavor to accommodate itself to the varying intensity of illumination. Moreover, a good artificial illuminant must be reliable. In other words, it must be able to furnish light without constant attention, and without danger of becoming accidentally extinguished.

Ordinary sunlight, as is well known, consists of a mixture of a great variety of colors. The color of natural bodies is entirely due to the light which falls on them. For example, a green leaf, illumined by sunshine, possesses the power of absorbing nearly all the sunlight but the green light, which it gives off. For such a green leaf to appear of its natural color by artificial light, this light must possess not only the exact tints of the various greens which it throws off in sunlight, but also the various proportions of such tints. The same is true for other colors. A good artificial illuminant, therefore, to be able to replace sunlight, should possess not only all the colors of the sunlight, but should also possess these colors in nearly the same relative intensities as does sunlight.

A requirement of an artificial illuminant, which is very difficult to fulfil, is that the light it produces shall not be localized, but shall uniformly illumine the spaces to be lighted. In order to meet this requirement, the artificial illuminant must readily yield itself to subdivision, that is, instead of giving a great amount of light at each of a few points, it should give a smaller quantity of light at a great number of separate points.

The light of all artificial illuminants is accompanied by heat, and, in nearly every case, the amount of heat so accompanying the light is very greatly in excess of what is essentially necessary. Such is true even in the case of sunlight. A notable exception, however, is found in the phosphorescent light of the firefly and the glow-worm, which yield light practically devoid of wasteful heat; *i. e.*, *non-luminous heat.*

Another requirement of an artificial illuminant is that it shall readily yield itself to control, that is to say, that it shall easily be turned on or off at a distance, otherwise, the location of the separate sources of light would necessarily be limited and thus prevent the most advantageous distribution. Finally, a good artificial illuminant should be readily adaptable to the purposes of decoration, or it will otherwise be shorn of many of its advantages.

The principal artificial illuminants in use at the present day, are, coal oils, gas, candles, arc and incandescent electric lamps.

The incandescent lamp is now so generally employed as an indoor artificial illuminant, and its advantages for this purpose are so evident, that it is scarcely necessary to show how much more fully it meets the requirements of a good illuminant than any of its predecessors. It suffices to say that it does not vitiate the air, is reliable, steady, can be maintained without any trouble on the part of the consumer, and readily yields itself both to sub-division and to decorative effects. When used on a comparatively large scale it can compete favorably as regards cost with any other artificial illuminant, and, even when employed on a small scale, its manifold advantages will frequently more than compensate for the disadvantage of a slightly increased cost.

As regards the ability of an incandescent lamp to produce a color of light akin to sunshine, it must be confessed that it is far from realizing this object, but this objection also exists to even a greater degree in nearly all other artificial illuminants. Coal oil, candles, gas, and oil lamps generally, do not produce a light so nearly akin to sunshine, as does the incandescent lamp. As we shall see later on the light emitted by an incandescent lamp can be made to approach more closely the characteristics of sunlight by increasing the temperature of the glowing carbon filament.

When the incandescent electric light was first introduced on a commercial scale, a belief existed that the extended introduction of this illuminant would be attended by many dangers, and it is true that, when negligently installed, such dangers do exist, but experience has amply proved that when installed with ordinary care, there is far less danger from the use of incandescent lighting than from the use of any other artificial illuminant.

So far as the safety of incandescent lighting for fire risk is concerned, a report of the Massachusetts Insurance Commissioner, giving data extending from 1884 to 1889, as to the origin of 12,935 conflagrations which took place in Massachusetts during that time, will speak favorably for this illuminant. Of this number of fires only 42 were attributed to electric wires. The breakage and explosion of kerosene oil lamps produced in the same time about 22 times as many fires, while the careless use of matches produced more than 10 times as many.

As to the danger to life, the evidence is still more in favor of electricity as an artificial illuminant. While it is true that fatal accidents do occur from contact with high-pressure wires, yet the proper installation of a system practically precludes the possibility of such accidents. Besides the immunity from fire which the electric lamp ensures, owing to the fact that the filament is sealed in a glass chamber, thus preventing contact with combustibles, it also possesses the marked advantage that

ARTIFICIAL ILLUMINATION. 15

it dispenses entirely with the use of dangerous friction matches, the lamp being readily capable of being lighted and extinguished at a distance by the mere turning of a switch. To thoroughly appreciate the danger to life from electricity, and to ascertain the extent to which it can be avoided, it will be necessary to understand some of the leading elementary principles of electrical science, and we will, therefore, postpone this consideration to a later chapter.

An extended experience with the incandescent lamp, has clearly established its advantages over gas or coal oil. Take, for example, convenience of night work, in a large textile manufactory. It is a well known fact that the air of such buildings, when illumined by gas, so rapidly heats and becomes so rapidly vitiated during summer time, that the amount and character of the work the workmen are capable of performing, necessarily suffers as compared with day work. Since the introduction of the incandescent lamp, however, the absence of increase of temperature and impure air, on prolonged runs, is so marked, that in many instances it has been found that the amount and character of the night work compares very favorably with day work during the summer season. This circumstance has frequently occasioned surprise to the public, since the temperature of the glowing filament in an incandescent lamp is quite high, but it must be remembered that while an incandescent lamps emits no gas, the filament not being consumed, a gas burner not only gives off all the gas that would escape from it were it unlighted, but, in addition, a much greater volume of heated air. Every cubic foot of ordinary illuminating gas requires for its combustion, the oxygen from about 20 cubic feet of air, so that a burner consuming 5 cubic feet per hour combines with the oxygen of about 100 cubic feet of air per hour. Besides vitiating such air by the products of combustion, and raising its temperature by the great amount of heat given off, this vitiation and increase of temperature requires much more thorough ventilation than when the incandescent lamp is employed. In cases where the character of night work requires keen sight, a highly heated vitiated air has an injurious influence upon the eye.

CHAPTER II.

EARLY HISTORY OF INCANDESCENT LIGHTING.

ELECTRICITY was first employed as an illuminant in the arc lamp. In this lamp, as is well known, two rods of carbon are first brought into contact, and then gradually separated while a powerful electric current is passing between them. A cloud of incandescent carbon vapor, called the *voltaic arc* is thus established, and the ends of the carbons, particularly that of the positive carbon, or the one from which the current flows, becomes intensely heated, forming a brilliant source of light.

The greatest difficulty attending the practical application of the arc system of lighting, arose from the fact that the arc lamp was too intense a source of light for most practical purposes within doors, and could not readily be subdivided into a number of smaller units; for, even if the space to be lighted would require all the light emitted by a single arc lamp, yet this light, coming from practically a single point, would necessitate the production of disagreeable shadows.

From very early times in the history of the application of electricity to lighting, the idea was conceived by various inventors of employing continuous conductors, instead of the discontinuous conductors in arc lighting. These continuous conductors were rendered incandescent by the passage through them of an electric current; or, in other words, the current heated them to a white heat. Various substances were employed for this purpose, at first in air, such as platinum or iridium wires, or other metals. These lamps, however, were not found to give practically useful results, since if the current strength was made sufficiently great to raise them to a white heat, although the light they would then emit would be quite satisfactory, yet disintegration would occur, from the free contact of the air, and would soon result in the destruction of the lamp. Moreover, the temperature at which the glowing wire becomes white hot, is so very nearly the temperature of its melting point, that any accidental increase in the current, even to a slight degree, would result in the fusion of the wire and the rupture of the lamp. If to avoid these difficulties, the lamp was burned at a lower temperature, the light it emitted was unsatisfactory, not only on account of its lessened intensity, but also because it was distinctly red and dull.

During the year 1878, a great improvement was effected in the platinum incandescent lamp, whereby platinum was obtained in a condition in which it was possible to employ safely much greater current strength without rapid deterioration. This process consisted essentially in sending a gradually increasing current of electricity through the wire while in a vacuous space. As is well known, platinum possesses in common with some other metals, the power, of absorbing or occluding within its mass, air or other gases. When such a wire is heated by the sudden passage of an electric current, this occluded gas is liberated explosively, and the wire thereby becomes cracked, fissured, and rapidly disintegrates. It was discovered that if the wire, while in a vacuous space, was subjected to the passage of a gradually increasing current of electricity, first beginning with a weak current, that the occluded gas was slowly liberated and that if, when a very high vacuum was obtained, the wire was maintained for a few moments at a temperature only a little below that of its melting point, it became physically changed, free from cracks and had its point of fusion raised. Moreover, the surface of the wire was altered, so that a given current strength produced a greater amount of light. The invention of a lamp consisting of such a treated platinum wire in a vacuous space was a marked step in the production of an artificial electric illuminant. Nevertheless, even the improved conductor did not answer commercial requirements, so that the improved platinum lamp did not come into any extended use.

To make the incandescent lamp practicable it was necessary to enclose the wire in a transparent chamber from which the air had been removed. Such an improvement would render the lamp more efficient for the following reasons:

(1) The absence of air would prevent the rapid disintegration of the wire.

(2) The absence of air would prevent the re-absorption or occlusion of air by the heated wire.

(3) The absence of air would prevent loss of heat, and much electric power would thereby be saved since a smaller current would bring the wire to the incandescent temperature.

(4) The absence of drafts of air would prevent irregular coolings and heatings of the wire, with consequent fluctuations of light.

(5). The wire would be protected from mechanical injury.

The early history of the art contains the names of numerous inventors who applied the foregoing principles for the purpose of producing satisfactory incandescent lamps. Without attempting to record in full all these early attempts, we will briefly allude to some of the more prominent of the early inventors in this field of artificial illumination.

In 1841, Frederick De Moleyn patented in England a process for the production of an incandescent lamp, based on the incandescence of platinum wire placed inside an enclosing glass chamber from which the air had been exhausted.

In 1849 Petrie produced an incandescent lamp in which short thin rods of iridium, or its alloys, were used as the incandescent material.

Considerable excitement was created in scientific circles in France, in 1858, by the announcement to the French Academy of Sciences by one of its members, of an improvement in incandescent lamps made by one De Changy. De Changy employed an incandescent platinum lamp of the form shown in Fig. 1, in which a wire of platinum G', is heated to incandescence by the passage of the current through it. The wire was enclosed in an exhausted glass vessel. The excitement caused by this lamp was due to the claim made by De Changy that he had solved the problem for the successful sub-division of the electric light. His lamp never came into commercial use.

The great improvement in lamps of this type was made by the substitution of carbon wires for platinum wires. The

26 ELECTRIC INCANDESCENT LIGHTING.

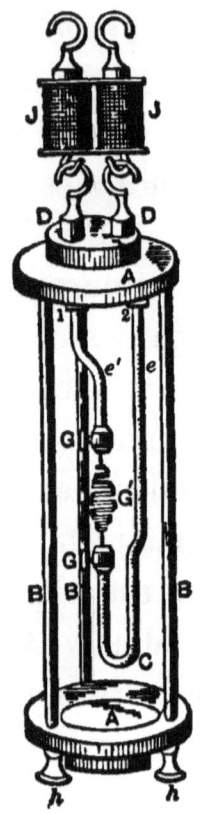

FIG. 1.—DE CHANGY'S LAMP.

honor of this discovery appears to be due to an American by the name of J. W. Starr, who employed plates of carbon placed inside a glass vessel containing a Toricellian vacuum. Starr associated himself with an Englishman of the name of King, and an English patent was obtained under the name of King. It is from this fact that the Starr lamp is not infrequently alluded to in literature as the King lamp, or sometimes, as the Starr-King lamp. Starr gave a successful demonstration in England with a candelabra of 26 of his lamps, the number of States then in the American Union. His untimely death, while on his return voyage to this country, retarded the progress of this invention. The Starr-King lamp is illustrated in Fig. 2, where A, is a rod of carbon clamped at its extremities to the rods C and D, and situated in a Torricellian vacuum above the surface of mercury in a barometer tube, the current passed through the rod A raising it to the incandescent temperature.

28 ELECTRIC INCANDESCENT LIGHTING.

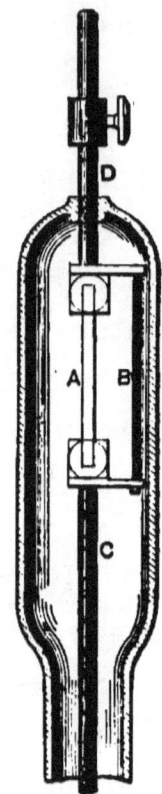

Fig. 2.—Starr-King Lamp.

The next invention worthy of notice was that of Lodyguine, who produced a carbon incandescent lamp in 1873. This invention was deemed by the St. Petersburg Academy of Sciences, of sufficient merit to warrant the award of a special prize. Lodyguine's lamp used needles of retort carbon terminating in blocks and placed inside an exhausted glass globe. In practice, in order to avoid the expense of exhausting the globe, Lodyguine sometimes left air within the globe and then sealed it hermetically, depending on the consumption of the residual air by the heated carbons to produce a space devoid of oxygen. This form of vacuum, however, was not found suitable for commercial purposes. Lodyguine's lamp was improved by Kosloff, who introduced modifications for the supports of the carbon pencils.

In 1875, Konn produced a lamp very similar to Lodyguine's. Konn's lamp is shown in Fig. 3. The glass globe B,

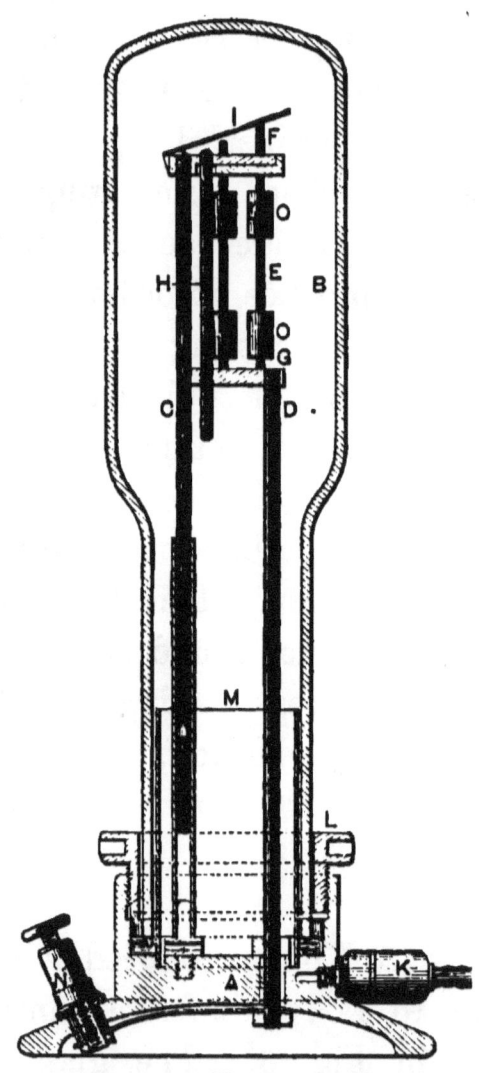

Fig. 3.—Konn's Lamp.

closed at the top, rests with its base upon a washer of soft rubber placed in the lamp base, and pressure is brought by the screw block L, in such a manner as to maintain an air-tight joint. The lamp was exhausted through the orifice K. The metallic base formed one terminal of the lamp, while the rod D, passing through the base A, in an insulating tube, formed the other terminal. Between the terminals F and D, were two rods of carbon, so arranged that one only was in circuit at any one time. The thinner central portion of these rods was heated to incandescence by the electric current passed through the lamp, while the thicker portions O, O, served to cool the extremities, at the points of contact with the supporting frame. After the first rod was consumed, it dropped out of position, and allowed the second rod to replace it. When

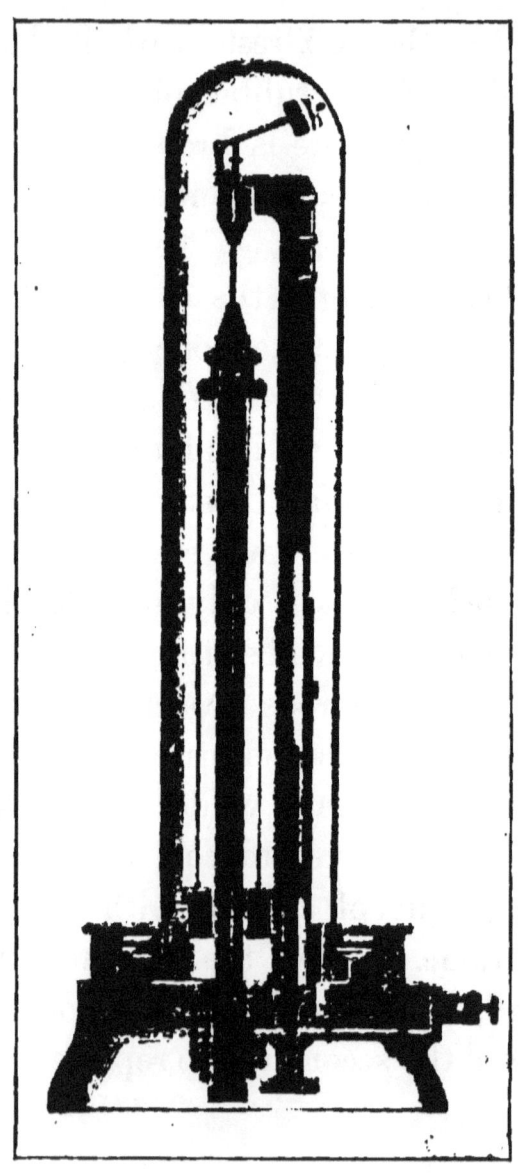

FIG. 4.—BOULIGUINE'S LAMP.

the second rod was consumed, the lamp became automatically short-circuited. The difficulty with this lamp was the too rapid consumption of the carbons.

In 1876 an improvement was made by Boulyguine, who produced a lamp intended to obviate this difficulty. In Boulyguine's lamp, the carbon is automatically fed upwards as it consumes away. A section of this lamp is shown in Fig. 4. Here the lower holder of the carbon rod has a slide in which guides press upwards under gravitational force, and feed the carbon filament up against the upper electrode as the short rod or filament is consumed.

Fig. 5, shows a form of Sawyer lamp. Here the light-giving medium consists of an incandescent carbon pencil at the top of the lamp. This consumes away slowly

34   ELECTRIC INCANDESCENT LIGHTING.

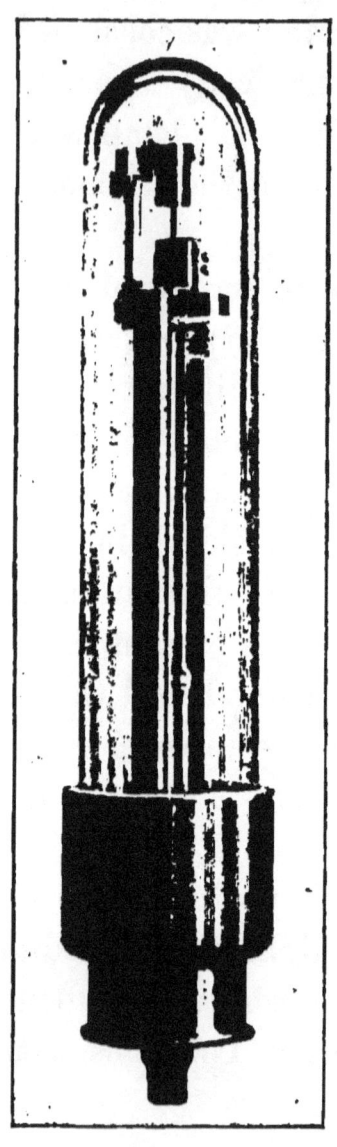

Fig. 5.—Sawyer's Lamp.

in operation at its contact with the upper carbon block, at the rate of about $\frac{1}{50}$th

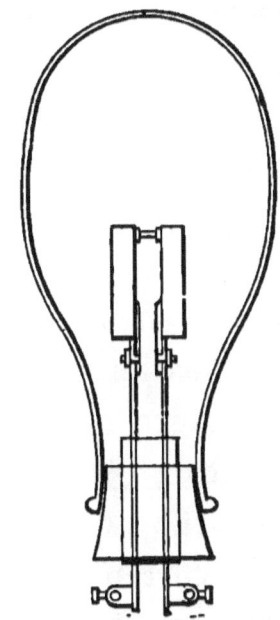

FIG. 6.—FARMER'S LAMP.

inch per hour. The carbon pencil was about eight inches in length and was forced upwards as consumption took place. The lamp was mounted in a glass globe partially exhausted.

Fig. 6, shows an early form of carbon lamp introduced by Farmer in 1879. Here a short horizontal carbon rod is gripped between two large metallic blocks in the exhausted globe.

All the lamps we have hitherto described have proved commercially useless, in the forms in which they were presented. What might have been the effect of introducing slight modifications into such lamps is beyond the province of this volume to consider. There can, however, be but little doubt that the want of a cheap source of electric current stood as much in the way of the progress of electric incandescent lighting, as any glaringly inherent difficulty in the structure of the lamps themselves.

Before leaving this brief history of the

early forms of incandescent lamps, it may be well to discuss some of the many forms that were devised for burning in the open air. These lamps strictly speaking were of a type which may be best described as *semi-incandescent lamps*. They operate essentially on the principle of sending a current through a slender rod of carbon pressed against a block or larger mass of the same material. Under these circumstances the slender rod is raised to intense incandescence, especially at its point of contact with the larger electrode, where, in reality, a miniature arc is formed. Various devices were employed in lamps of this type for feeding the carbon rod or pencil as it was gradually consumed. In some lamps an enclosing glass chamber was employed in order to reduce the consumption of the carbon as far as possible. This type of lamp passes insensibly into lamps of the

38 ELECTRIC INCANDESCENT LIGHTING.

purely incandescent form. In fact, as will be observed, some of the preceding lamps were of the semi-incandescent type.

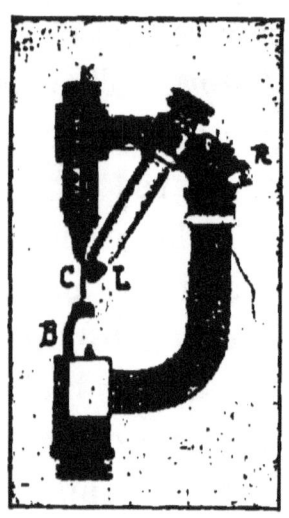

FIG. 7.—REYNIER'S LAMP.

One of the most successful of the early lamps of the semi-incandescent type was that of Reynier. In the Reynier lamp a movable rod of carbon C, Fig. 7, of small diameter, supported as shown, rests against a contact block *B*, of graphite. In order to limit the incandescence of the rod to its lower extremity, a lateral contact piece *L*,

FIG. 8.—REYNIER'S LAMP.

is kept pressed against the rod *C*, by means of a spring *R*. The current, therefore, passes through the lateral contact piece *L*, through the slender rod to the contact block of graphite *B*, so that only the portion of the rod between these points is rendered incandescent, by far the greatest amount of light coming from the tip or extremity *J*. Fig. 8, is a semi-diagrammatic view of the same lamp arranged, however, to be used in connection with an external globe. The two binding posts at the top of the lamp form its terminals. No vacuum is necessary with this form of lamp since the slender rod of carbon is fed downwards as the point consumes away.

In 1878 an Englishman, named Werdermann, took out a patent for a lamp founded on a somewhat similar principle. In the Werdermann lamp, shown in Fig. 9, as in the Reynier lamp, means were devised for pressing a slender carbon rod against a fixed electrode, and feeding this rod upwards as it consumed away. Werdermann employed for his fixed electrode a carbon disc and advanced a slender carbon rod by the action of a weight or counterpoise. In order to avoid the casting of shadows downwards, a notable defect in the

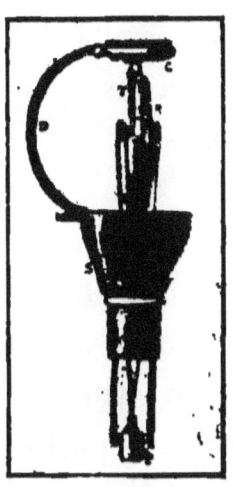

Fig. 9.—Werdermann's Lamp.

Reynier lamp, Werdermann inverted his lamps, as shown in the figure. Lamps of the combined Reynier-Werdermann type at one time were in fairly successful practical use, and formed one of the features of the Electrical Exhibition of 1881˚ at Paris.

The advent of the really successful incandescent lamp dates from about 1879, and from this year, the growth of the incandescent electric lamp industry has been extremely rapid. The year 1879, therefore, marks the entrance to the epoch of contemporaneous history, into which we are unable to enter at the present time. We will, therefore, close this extremely brief history of the art, referring the reader for further particulars to contemporaneous literature.

CHAPTER III.

ELEMENTARY ELECTRICAL PRINCIPLES.

THE light emitted by the glowing filament of an incandescent lamp is one of the effects produced by the passage of the electric current through it. Before proceeding to a discussion of the operation of the incandescent electric lamp, it will be necessary to consider briefly the leading elementary principles concerning the production and flow of electricity.

An electric flow is always produced by the action of what is termed *electromotive force*, generally contracted E. M. F. No electric source produces electricity directly.

What is produced is an electromotive force, and an E. M. F., in its turn, produces an electric current when permitted to do so. Thus, in such an electric source as a voltaic battery, an E. M. F. is produced, and this independently of whether it is permitted to establish an electric current or not. If an E. M. F. be permitted to act, it will invariably establish an electric current. In order to do this a conducting path must be opened to it. Such a conducting path is called *a circuit*.

For convenience, it is universally agreed to regard electricity as leaving an electric source at the point called the *positive terminal* or *positive pole*, and re-entering the source, after having passed through the circuit, at a point called the *negative terminal* or *negative pole;* that is to say, electricity is regarded as flowing from the positive to the negative pole, in the external portion of the circuit, and from the negative to the positive pole, in the internal portion of the circuit, that is within the source.

Since all electric currents are due in the first instance to the action of an E. M. F., it is necessary to obtain definite ideas concerning this force. E. M. F. is to electric flow the analogue of ordinary pressure to the flow of liquids or gases; that is to say, a flow or current never occurs in a liquid or gas, except as the result of difference of pressure acting on it, and the liquid or gaseous flow is always directed from the point of higher pressure to the point of lower pressure. So in the electric circuit, a current of electricity never flows unless there be a difference of *electric pressure*, or an E. M. F., and the electric flow is always directed from the point of higher to the point of lower electric pressure.

E. M. F. is measured in units called *volts*. Thus, a voltaic cell, of the type known as a Leclanché cell, has an E. M. F. of, approximately, 1 1/2 volts. Such a cell is shown in Fig. 10. A carbon plate CCC, is connected to the positive terminal P, while a rod of zinc ZZ, is connected to the negative terminal N. This cell produces, when in good order, an E. M. F. of about 1 1/2 volts, even when on *open circuit*, that is when the terminals P, and N, are disconnected as shown, and are, therefore, producing no current. If, however, these terminals be connected through a conducting path, or circuit, a current of electricity will flow under the pressure of 1 1/2 volts, from the positive terminal P, through the external portion of the circuit back to the negative pole *N*, and from the

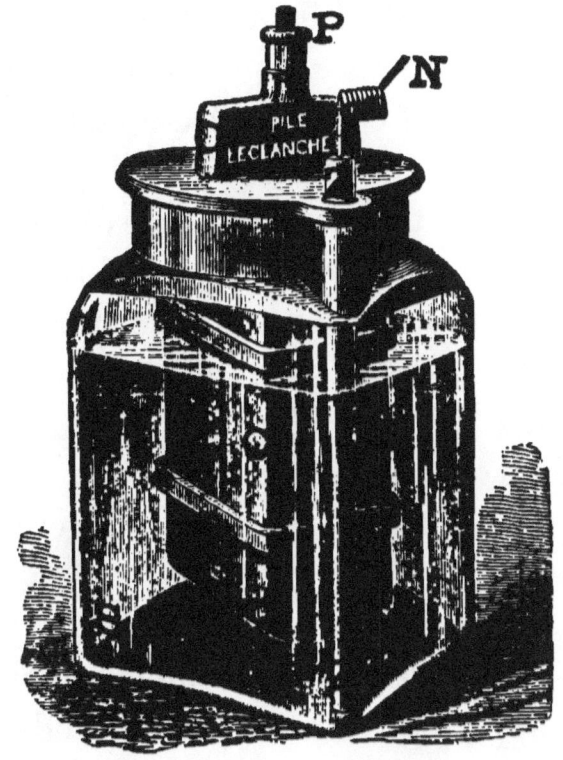

Fig. 10.—Leclanché Cell.

zinc plate through the solution *S S*, to the positive terminal, thus completing a circuital path.

When an E. M. F. greater than 1 1/2 volts is required, from cells of this type, a number of such cells are so connected together as to act as a single source. Such a combination of cells is called a *battery*. If, for example, two such cells be joined together, with the negative pole of one cell connected to the positive pole of the other, they would produce a battery of three volts E. M. F., and 100 such cells connected in this way, or connected *in series*, as it is called, would produce a battery having a total E. M. F. of approximately 150 volts.

Voltaic batteries are seldom employed for supplying current to incandescent lamps except on a small scale, for the reason that the cost of the current so produced would be excessive. In almost all cases of electric lighting, the E. M. F. is obtained from a *dynamo-electric generator*, a machine for producing electric power by the expenditure of mechanical power. Dynamo-electric machines, suitable for incandescent lighting, will be considered in a subsequent chapter.

When an E. M. F. is permitted to act upon a closed conducting path or circuit, the value of the current strength produced therein, that is, the amount of electricity which flows in a given time, will depend upon two circumstances; namely,

(1) Upon the value of the E. M. F.; *i. e.*, the number of volts.

(2) Upon a property of the circuit called its *electric resistance*, that is to say, the opposition it offers to the flow or passage of electricity through it. Thus, if the cell shown in Fig. 10 has an E. M. F. of 1 1/2 volts, and produces a much

greater current strength in one circuit than in another, it is because the latter circuit offers a greater resistance to the flow of electricity than the former. Some idea of the action of resistance, in opposing the flow of electricity in an electric circuit, can be had by considering the analogous case of the flow of gas through a pipe or main. A long narrow pipe will obviously offer a greater resistance to the passage of gas through it, from a reservoir, than a larger short pipe.

The resistance of an electric circuit is measured in *units of electric resistance* called *ohms*. The ohm is a resistance such as is offered by about two miles of ordinary overhead trolley wire, or about one foot of very fine copper wire, No. 40 A. W. G., which has a diameter of about $\frac{3}{1,000}$ th inch.

ELEMENTARY ELECTRICAL PRINCIPLES.   51

The electric resistance of a circuit is a very important quantity. The following resistances will, therefore, be of interest.

An ordinary Bell telephone receiver has a resistance of about 75 ohms.

An ordinary telegraph sounder has a resistance of about 2 ohms.

An ordinary incandescent lamp, of 16 candle-power, intended for 115-volt circuits, has a resistance, when lighted, of about 250 ohms.

The electric resistance of a conductor depends upon its length, its cross-sectional area, and upon the nature of the material of which it is composed. The resistance increases directly with the length of conductor, and inversely as the cross-sectional area. Thus, if a mile of trolley wire, weighing 1,690 lbs. has a resistance of 1/2 ohm, two miles will have a resistance of one ohm, and 10 miles, a resistance of 5 ohms. If the above trolley wire be doubled in cross-sectional area, and, consequently, in weight, so as to weigh 3,380 lbs. per mile, its resistance will be only 1/4 ohm per mile, or 2 1/2 ohms in 10 miles.

The resistance of wires, each a mile long, and of the same diameter as ordinary trolley wire, (0.325″) would be different with different materials. Thus, while such a wire of copper, has a resistance of about 1/2 an ohm, an iron wire of the same size, length and diameter, would have a resistance about 6 1/2 times greater, or about 3 1/4 ohms, and when of lead, about 12 times greater, or about 6 ohms. Consequently, in order to compare the relative resistance of wires of different materials, it is necessary to refer each material to a common standard of dimensions. This is done by considering the resistance of a wire having unit length and unit cross-sectional area, that is to say, the resistance of a wire having a length of one centimetre and a cross-sectional area of one square centimetre. The resistance of a wire with such unit dimensions is called its *specific resistance* or its *resistivity*. The resistivity of standard soft copper is 1.594 *microhms;* *i.e.*, 1.594 millionths of one ohm. If then a wire has a length of one kilometre (100,000 centimetres) and a cross-sectional area of 1 square centimetre, it would have a resistance of 1.594 × 100,000 microhms = $\frac{1.594 \times 100{,}000}{1{,}000{,}000}$ = 0.1594 ohm, at the temperature of melting ice.

The resistivity of a wire is the scientific standard for comparing its resistance with that of other wires of the same length and cross-sectional area. In English-speaking countries, where the centimetre is not in general use as the unit of length, the standard called the *circular-mil-foot* is very commonly employed. A *circular-mil* is the area of a wire one mil, or $\frac{1}{1,000}$th of an inch, in diameter. This is not to be confused with the area of such wire expressed in square inches. The number of circular mils cross-sectional area in any wire, is obtained by squaring its diameter in mils. Thus a wire half an inch in diameter, would be a wire of 500 mils diameter, and the number of circular mils cross-sectional area in such wire would be 500 × 500 = 250,000. Such a wire would have a resistance per foot of $\frac{10.35}{250,000}$ = 0.000,00414 ohm. A circular-mil-foot of standard soft copper at 20° C. has a resistance of 10.35 ohms.

The resistivity of a material varies with its temperature. In the case of most metallic substances, the resistivity increases as the temperature increases. Thus, the resistivity of copper wire is about forty-two per cent. greater at the boiling point of water, than at its freezing point; or, a copper wire would have about forty-two per cent. more resistance, at the temperature of the boiling point of water, than at its freezing point. The resistivity of insulating materials, however, diminishes as the temperature increases. Carbon behaves in this respect like an insulating material, its resistivity diminishing as its temperature increases. An ordinary incandescent lamp has about twice as much resistance when cold, as it has when heated by the electric current to an incandescent temperature.

The current strength, which passes

through any circuit, is measured in *units of electric flow* called *amperes*. As in the case of a current or flow of gas, we may estimate the flow as so many cubic feet of gas per minute, or per second, so an electric flow, may be estimated as so many units of electric quantity per second. The *unit of electric quantity* is called the *coulomb*, and is the quantity of electricity, which would flow in one second through a circuit having a total resistance of one ohm, when under a pressure of one volt. A rate of flow equal to one *coulomb-per-second* is the unit of electric flow or current, and is called the ampere. A current of 10 amperes, therefore, means a flow, or transfer, of 10 coulombs of electricity in each second. The ordinary incandescent lamp of 16 candle-power, operated at a pressure of 115 volts, requires to be supplied with a current of about 1/2 ampere.

The relations existing in any circuit under a given resistance and E. M. F. are readily determined by reference to a law discovered by Dr. Ohm, and named *Ohm's Law*. This law may be expressed briefly as follows:

The current strength in any circuit is directly proportional to the total E. M. F. acting in the circuit, and inversely proportional to the total resistance in the circuit.

If the pressure or E. M. F. be measured in volts, and the resistance in ohms, the current strength that passes in amperes may be briefly expressed as follows:

$$\text{Amperes} = \frac{\text{Volts}}{\text{Ohms}}$$

For example, if a circuit comprise a dynamo and an external path consisting of lamps and wires, so that the E. M. F. in the dynamo is 100 volts, and the resistance of the circuit is 2 ohms, then the current strength passing through the circuit will be $\frac{100}{2} = 50$ amperes.

Again, if an incandescent lamp has a resistance (hot) of 220 ohms, and is connected to a pair of mains between which the electric pressure is steadily maintained under all circumstances at 110 volts, then the current, which will pass through the lamp from the mains will be $\frac{110}{220} = \frac{1}{2}$ ampere.

We have already alluded to the fact that a current never flows in water unless a difference of pressure exists therein. In order to produce this difference of pressure, the water has usually to be raised to a higher level; and, to do this, energy is required to be expended, or work performed, on the water. The same is true of electricity. In order to produce an electric flow, energy requires to be expended, or work produced, by the electric source.

Work is never done unless force acts through a distance. A force that is merely producing a pressure on a body, but no motion of the body, is naturally performing no work. When, for example, a block of granite is raised through a vertical distance against the earth's gravitational force, work is done. The amount of such work is measured by the force which acts, multiplied by the distance through which it acts. For example, if a block weighing 200 pounds be raised through 10 feet, an amount of work will be done expressed in a common *unit of work* called the *foot-pound*, as being equal to 200 pounds × 10 feet = 2,000 foot-pounds.

If this weight when raised be placed on a shelf or other support, it would still produce a pressure of 200 pounds upon the support, but would be doing no work.

Another unit of work is called the *joule*, and is approximately equal to 0.738 foot-pound, so that a foot-pound is about thirty-five per cent. greater than a joule, or 1 foot-pound = 1.356 joules. A man weighing 150 pounds, and raising his weight through a distance of 100 feet by walking upstairs, necessarily performs an amount of work against gravitational force equal to 100 × 150 = 15,000 foot-pounds = 20,340 joules.

When an electric current passes through a circuit, work is always done by the E. M. F. which drives the current through the circuit. The amount of work done by the E. M. F. is expressed in joules, as the product of the pressure and the quantity of electricity it drives. Thus, if a quantity of electricity equal to 100 coulombs, passes through an electric circuit under a pressure of 50 volts, then the amount of work which will be expended in the passage, will be $50 \times 100 = 5{,}000$ joules $= 3{,}690$ foot-pounds. A joule is, therefore, equal to a *volt-coulomb*. This work will always be taken from the source of the driving E. M. F. Thus if the dynamo in the last example, supplied the E. M. F., then an amount of work equal to 5,000 joules will have been taken from the dynamo, and this amount of work must have been delivered to the dynamo through its driving belt. Or, if the E. M. F. had been supplied by a voltaic battery, this amount of work would have been supplied at the expense of the zinc and the liquid in the battery; that is to say a certain amount of zinc would have been consumed, thereby liberating at least 5,000 joules of work.

It is necessary to distinguish carefully between the amount of work performed in any case and the rate at which such work is performed; for example, so far as the result attained is concerned, the same amount of work is done, when the man before alluded to, weighing 150 pounds, raises himself through a height of 100 feet by the stairs, whether he performs this work in one minute or in ten minutes, but his *rate-of-doing-work*, or his *activity* would be 10 times greater in the former than in the latter case. In the former, his activity would be 15,000 foot-pounds-per-minute, or 250 foot-pounds-per-second, while in the latter case it would be only 1,500 foot-pounds-per-minute, or 25 foot-pounds-per-second. Activity is commonly rated either in terms of a *unit of activity* called a *foot-pound-per-second*, or in a unit called a *horse-power*, one horse-power being equal to 550 foot-pounds-per-second, or 33,000 foot-pounds-per-minute. Thus, in the preceding case, the man would be expending an activity of $\frac{250}{550} = 0.455$ horse-power in the first case, and 0.0455 horse-power in the second case.

Electric activities are similarly measured in *units of electric power* or *activity* called *watts*, the watt being equal to an activity of a *joule-per-second*, or a *volt-coulomb-per-second*, or a *volt-ampere*.

Thus if a current of 1/2 ampere passes

through an incandescent lamp under a pressure of 115 volts, the electric activity or rate of expending energy in the lamp, will be $115 \times 1/2 = 57.5$ watts, or joules-per-second $= 42.4$ foot-pounds-per-second.

CHAPTER IV.

PHYSICS OF THE INCANDESCENT ELECTRIC LAMP.

BEFORE proceeding to a detailed description of the incandescent lamp and the methods employed in its manufacture, it will be advisable to obtain a general insight into the physical laws controlling its operation.

Briefly speaking, the operation of the incandescent electric lamp is based upon the principle of raising the temperature of a thin thread or filament of some refractory substance, such as carbon, as far as is consistent with practical working. In order to avoid the oxidation of the carbon filament, it is enclosed in a glass lamp bulb in which a vacuum is maintained. Briefly, the function of the electric lamp is to convert electric energy economically into luminous energy or light.

Light may be defined in two distinct senses.

First, subjectively, as the physiological effect produced through the eye on the mind by the radiations emitted by a luminous body. In this sense, lights differ in their color and intensity; that is, we are conscious of different perceptions both of color and of intensity.

Second, objectively, as the physical cause which produces the sensation of light. In this sense light can have an existence independent of the eye. Objectively, light consists of rapid vibrations or to-and-fro motions in a medium called the universal or luminiferous ether, which permeates all substances and spaces.

The luminiferous ether is believed to be an extremely tenuous and highly elastic medium, which not only fills interstellar space, but which even permeates the densest forms of matter. In this sense any vibrations, or to-and-fro motions, of the luminiferous ether, even though not capable of affecting the eye, are properly spoken of as light, but of course the light, which it is the function of the incandescent lamp to produce for the purposes of illumination, must necessarily be light in the physiological sense.

The rapidity of the oscillations in the ether which constitute light is usually enormously great. This frequency has been indirectly measured up to over 800 trillions; *i. e.*, over 800,000,000,000,000 double vibrations per second, and they may exist down to comparatively low frequencies, although they have only been measured in heat radiation as low as about 100 trillions per second. Those frequencies which lie between 390 trillions and 760 trillions, are capable of affecting the normal eye as physiological light.

All bodies emit from their surfaces radiations or waves into surrounding space. A body at ordinary temperatures emits only waves of a comparatively low frequency, far below that which is capable of affecting the eye physiologically as light. As the temperature of the body is increased, not only does it emit these waves of low frequency more powerfully, but, in addition, it also emits waves of a higher frequency. When a temperature of about 500° C. is reached, the highest frequency emitted by the body will be just about 390 trillions of double vibrations, or complete to-and-fro motions, per second, and will be just visible to the eye. The body is then said to be at the temperature of dull red. As the temperature is still further increased, the total radiation of waves from the surface increases, and at the same time higher frequencies are introduced. These affect the eye successively as orange, yellow, green, blue, indigo and violet; and, finally, all these colors being present, the body is said to be white hot, and has a temperature of roughly 1,500° C. If the temperature be still further increased, the *luminous intensity; i. e.,* the amount of visible radiation per unit area of its surface, will increase, and at the same time still higher frequencies will be introduced into these vibrations, which are necessarily invisible to the eye. These are called *ultra-violet rays*, or those rays above the violet.

If we analyze sunlight, we will find that it contains all frequencies from the lowest that we can measure up to the ultra-violet. The greatest intensity of these vibrations falls within the limits of the invisible frequencies, only about thirty per cent. of the vibrations which we receive at the earth's surface being capable of affecting the eye. The visible frequencies combine to produce on the eye an impression, known as white light. In other words we judge a luminous body as white, when it emits frequencies of the character and general proportions which exist in sunlight, or the relative proportions of red, green, yellow, and blue frequencies or vibrations will be the same in this light as in sunlight.

Generally speaking, nearly all sources of artificial light emit frequencies which are the same as those in sunlight, but the amount of radiation in the different frequencies or colors varies markedly from sunlight. Thus a candle while emitting all the visible frequencies of sunlight has a marked preponderance of red rays relative to the number of violet rays, and consequently has a reddish yellow tint.

The color of a body depends upon two things; viz., first upon the quality of the light it receives, that is upon the proportionate distribution of the various frequencies, and second upon the selective properties of its surface. When light falls on a colored body, the surface of the body possesses the power of absorbing some of the frequencies and throwing off others unabsorbed. Thus, a blue body, illumined by sunlight, absorbs practically all the frequencies except those of the blues, which it emits. For the blue body to be visible in its proper tint, therefore, it is necessary that the light which illumines it shall not only contain blues, but shall also contain the same proportionate quantities of the different tints of blue as sunlight. If these colors be not present in the artificial light, such a blue body would fail to possess its characteristic daylight color-value. Consequently, an artificial illuminant, in order to replace sunlight for the purpose of revealing the proper daylight color-values of bodies, must contain not merely the same frequencies as sunlight, but also the same relative distribution of such frequencies.

All light serves either to distinguish the form and shape of bodies, or their colors. For the latter, as we have seen, an approximate imitation of sunlight is necessary; for the former, almost any frequency of light will be sufficient. For example, if differently colored bodies be observed in a dark room, by means of an artificial light containing practically but one frequency, and called usually a *monochromatic light*, only the color of this light will be visible, all other colored objects will appear black, or devoid of color. A yellow, monochromatic light, may be obtained, for example, by the burning of alcohol on a wick soaked in common salt. In the pure yellow light this emits, all yellow colored objects will appear in nearly their true tints, while the reds, blues, greens, and other tints will appear devoid of color. All objects, however, whatever their color, will show their form even when so illumined.

The light emitted by an incandescent electric lamp is not capable of giving true sunlight color-values. An analysis of the light from the glowing filament, shows a relative preponderance of red and yellow rays and a deficiency of the blue and violet, or high-frequency waves. This is owing to the fact that the temperature of the glowing filament cannot be raised sufficiently high to emit the sunlight proportions of the higher frequencies. Consequently, reds and yellows, viewed by incandescent lamp light, give nearer approach to their sunlight values than other colors. It is true that by raising the temperature of the carbon filament we obtain a closer approach to sunlight radiation, but, at the same time, for reasons

PHYSICS OF INCANDESCENT LAMP. 75

which will be subsequently explained, the life of the lamp is greatly shortened.

A hot body loses its heat in one of four ways; viz., by radiation, by conduction, by convection and by molecular transfer.

Were the glowing filament in an absolutely vacuous space, it is evident that being supported on a comparatively slender base, apart from the trifling loss of heat through this base or support there would be no other means for losing the heat save by radiation. Once admitting within the lamp chamber even a minute trace of gas such as would make the pressure one-millionth part of that in the atmosphere, then pure radiation would cease to form the sole means of losing heat, and molecular transfer would begin to act. That is to say the air molecules coming into contact with the heated filament would be shot off from its surface in straight paths, and, in a highly attenuated atmosphere, the molecules would nearly all fly to the chamber walls without mutual collision. Heat energy is required to produce this motion, and the loss of energy so occasioned forms an additional means whereby a heated body in a rarified atmosphere parts with its heat.

As more air is admitted to the interior of an exhausted lamp, the collisions of the air molecules, flying from the heated surface, become more frequent, and the frequency with which they are returned to the hot surface also increases, increasing thereby the loss of heat by molecular transfer. As soon, however, as the *mean free paths*, or uncollided path, of the molecules becomes a small fraction of the distance between the filament and the wall of the chamber, the frequency with which the molecules return to be shot off from the hot surface increases very slowly, so that beyond this point there is scarcely any increase in loss of heat by molecular transfer as air is admitted into the chamber. As more air is gradually admitted into the chamber, however, the loss by simple convection increases, *i. e.*, by a thermal stirring of the air owing to differences of density, or local winds, within the lamp chamber.

The radiation emitted by the ordinary incandescent electric lamp is principally non-luminous; that is to say the greater portion possesses a frequency below the inferior limit of visibility. An ordinary 16-candle-power lamp has an activity, when in operation, of about 50 watts. Of this activity about 48 watts are expended in non-luminous radiation, and only about 2 watts, or four per cent., in luminous radiation. The problem that still remains to be solved in the incandescent lamp, as it does indeed in the case of all artificial illuminants, is to produce a radiation which shall lie wholly, or almost wholly, between the limits of visible frequencies. As it exists to-day, in the case of the incandescent lamp, the energy is expended in producing about ninety-six per cent. of objectionable heat radiation, and four per cent. of light radiation. Even this, however, has a luminous efficiency superior to that of gas and oil.

It has been found that the light emitted by the firefly and the glow-worm is practically confined to the visible limits of frequency. Could an incandescent lamp be made to restrict its radiation to such frequencies, the problem of cheap light might be solved. Unless, however, such light could be made to possess these frequencies in sunlight proportions, it is questionable whether the problem of an efficient illuminant would be solved from the color standpoint.

Passing by the question of the production of lamps which shall yield frequencies characteristic of the light of the firefly and glow-worm, there would appear to remain but one direction in which the same result might be reached; namely, by obtaining a substance which possessed marked powers of *selective radiation;* that is, a high ability to radiate waves of high frequency, and but little for radiating those of low frequency.

When an electric current is sent through

the filament of an incandescent lamp, the activity expended in the lamp will be the product of the pressure at the lamp terminals in volts, and the current, in amperes, passing between them. This activity will be entirely expended as heat in the substance of the filament, and this heat will be liberated from the surface in virtue of the increase of temperature of that surface. As the amount of heat liberated in the filament increases; *i. e.*, as the current strength passing through it increases, the temperature of the filament is compelled to rise in order to emit the activity which is developed in its mass. If the surface of the filament be large, it will take a large total amount of activity in the lamp to maintain a large radiation per square inch, or per square centimetre; whereas, if the surface of the filament be small, the opposite result will be produced. Consequently, a certain relation must exist between the surface, the length, and the cross section of the filament, in order that, at the pressure intended for the circuit, the radiation per square inch, or per square centimetre, shall be sufficient to bring the lamp to the proper temperature. The nature of the surface of the filament determines, to a great extent, the character and amount of the radiation which it will emit. It might be supposed that the same activity per square centimetre of surface would be attended by the same temperature and relative distribution of vibration frequencies. Such, however, is not the case, the character of the surface having a marked influence upon its emissivity, one type of carbon producing, with a given activity of surface, a greater amount of light than another.

The temperature at which ordinary incandescent lamp filaments are operated, is estimated at about 1,345° C. If this temperature be exceeded, by only a few degrees, although the candle-power of the filament will be materially increased, yet the disintegration of the carbon, by a process akin to evaporation, will be rapidly brought about. Thus, an increase of 2° C. is believed to be accompanied by an increased candle-power of about three per cent., but this gain is at a marked decrease in the life of the lamp.

CHAPTER V.

MANUFACTURE OF INCANDESCENT LAMPS. PREPARATION AND CARBONIZATION OF THE FILAMENT.

The most important step to be taken in the manufacture of an incandescent electric lamp is the preparation of the filament. Of all substances which have been employed for filaments, carbon alone has been found to meet the requirements of use. In the first place, carbon is highly refractory; that is, capable of withstanding a high temperature before reaching its point of volatilization. In the next place, its resistivity is high, so that a high resistance can be readily given to a short length of filament, with a correspondingly high electric pressure at the terminals, for a given activity in the lamp.

Carbon is universally employed for lamp filaments, not only for the reasons above pointed out, but because this material readily lends itself to being fashioned and shaped into the filament, prior to being subjected to various processes of carbonization, intended to ensure a nearly pure form of hard high-resisting carbon.

A great variety of materials have been employed for producing the carbon filaments of incandescent lamps. All these substances agree in that they are of such a nature as will yield a nearly pure carbon when subjected to *carbonization, i. e.*, to the action of heat while out of contact with air. Substances suitable for this purpose may be divided into two sharply marked classes; namely,

(1) Carbons of fibrous origin, such as bamboo.

(2) Structureless carbons, such as pastes or mixtures of finely ground carbon incorporated with some suitable carbonizable liquid.

Among substances of fibrous origin, that have been employed for incandescent filaments, may be mentioned paper, bamboo, bass fibre, cotton thread, and silk thread, both of the latter substances being first subjected to a process which is called the *parchmentizing process,* by treatment with sulphuric acid. Cellulose, treated in such a manner as to be converted into a variety of gun-cotton and subsequently carbonized, has also been employed. This material, although of fibrous origin, would by its treatment be rendered capable of classification as structureless material. Without here entering into a full description of the various processes required for the manufacture of filaments from the above materials, it will suffice to describe in detail the process adopted in a few cases.

In the manufacture of a bamboo filament, carefully selected bamboo is employed, from which both the softer portions in the interior, and the hard silicious parts near the surface, have been removed. The material is then cut or fashioned, by the aid of a cutting tool, into filamentary strips, care being taken to obtain as nearly as possible the same area of cross section by passing the filaments through gauges. The filaments so prepared are then converted into hard carbon by means of a carbonizing process.

PREPARATION OF THE FILAMENT.

In the production of a filament from loosely spun pure cotton, the thread is first cleansed from grease by boiling in soda or ammonia, this cleansing being necessary for ensuring uniformity of action on the part of the sulphuric acid used in a subsequent process. The thread is then thoroughly washed in water and afterward soaked in sulphuric acid of specific gravity 1.64. The time of immersion is exceedingly short, varying with the thickness and character of the thread, from three to fifteen seconds. On removal from the acid, the thread is again washed in water. After removal from the water, care must be taken to avoid warping. The dried thread in this condition has the appearance of catgut and is called *amyloid*.

The parchmentized thread has a rough surface, and is too irregular in diameter to permit it to be subjected in this condition to the carbonizing process. In order to ensure uniformity in its diameter, and smoothness of its surface, it is passed through a series of draw plates, until a sufficient amount has been removed from it to permit the cutting tool to act on all portions of its surface. These draw plates are made either of steel, or of jewels, the latter being the most frequently employed. The thread so produced is then subjected to the carbonizing process.

Another process, which also produces an amyloid thread from pure cellulose, is sometimes employed. Here pure cellulose is dissolved at a temperature of about the boiling point of water, in zinc chloride of specific gravity 1.8. The viscous mass so obtained is then forced or squirted under pressure through a die of suitable diameter

PREPARATION OF THE FILAMENT.

into a vessel containing alcohol, which causes a hardening of the filament. If this process is properly carried on it produces a filament of uniform cross-sectional area, so as to dispense with the necessity for passing the thread through shaving dies. Care has to be taken to avoid the formation of air bubbles in the viscous mass, which would either result in a breaking of the thread, or in a lack of homogeneity of the filament. This process is now in very extensive use.

Another process has been invented for the production of an artificial material suitable for cutting or shaping into filaments for incandescent lamps, and subsequent carbonization. This process consists essentially in converting cellulose, obtained from cotton, into a substance known as pyroxyline, or gun-cotton, whence it is converted into *celluloid*. The celluloid is rolled into thin sheets. The sheets in this condition are not yet fit to be subjected to the carbonizing process, and must first be again converted into cellulose. This change is effected by treating them with ammonium sulphide, which produces structureless, cellulose sheets. These are then treated with bisulphide of carbon, or turpentine, to remove all traces of sulphur, and are then ready to be cut or shaped into filaments and carbonized.

The filaments so cut or shaped by the preceding processes, must be subjected to a carbonizing process. During carbonization a number of changes occur in the filament. In the first place the filament becomes hard and brittle, and, to a certain extent, acquires a definite shape or form, so that it is necessary before carbonizing it to give it the form which it is to permanently retain. In the next place the filament shrinks perceptibly during carbonization, so that means must be devised for permitting this shrinking to take place without rupture.

The means employed for the carbonization of the filament, will vary with the shape and character of the material of which it is made. The degree and duration of the heat employed in carbonization will also vary with the character and method of preparation of the material. It will, therefore, be necessary briefly to describe the methods employed for the carbonization of particular characters of filaments. In this connection, however, it must be remarked that the manufacture of the incandescent lamp, as it is carried out to-day, is, to a great extent, a secret process. Therefore, such descriptions must necessarily be limited to known processes which have been tried and which have been found successful in practice.

In the case of the bamboo filaments, the material, shaped as already described, is placed in a suitable box or receptacle formed of carbon, or other refractory substance, under such conditions that the filament shall be permitted to contract freely while being subjected to a constant and even tension. In Fig. 11, such a box is shown with the filaments to be carbonized placed in position. This box is in two parts as shown, an outer air-tight chamber, and an inner forming plate. The filaments to be carbonized are placed in the inner groove around the block EH, with

PREPARATION OF THE FILAMENT.

their extremities secured beneath the bridge piece *a h*, which is fixed in an outer frame. As the filaments contract

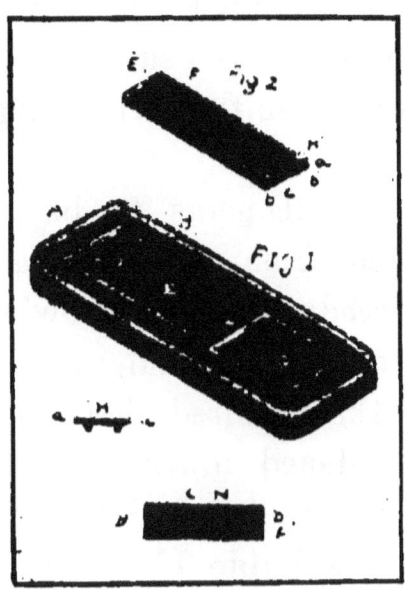

FIG. 11.—CARBONIZING BOX.

during the process, since they cannot draw up into the groove from the bridge, they pull the whole block *E H*, forward under the bridge, and so maintain a steady and uniform tension upon the filaments. A suitable cover, provided with a flange fitting into the outer groove, is placed on the box. A number of such boxes are packed together into a suitably closed flask which is placed in the carbonizing furnace.

When the filaments to be carbonized are prepared from cotton thread, by the process already described, owing to the pliability of the material, it is necessary to give it the required shape before it is set or hardened during carbonization. This is accomplished by winding the thread on a suitable block or form. As in the case of the bamboo filament, means must be provided for allowing shrinkage to take place. This is accomplished by means of a *carbonizing frame*, which consists essentially of a suitably shaped block and an end piece of carbon, connected to-

PREPARATION OF THE FILAMENT.

gether by sticks of wood. These sticks are firmly fixed to the end piece and pass loosely into openings in the main block which rests on them. The threads to be carbonized, are wrapped tightly around the block and end piece, which is so placed in the carbonizing box that on the shrinkage of the thread, the wooden sticks, whose dimensions have been carefully selected, shrink to the same extent, thus permitting the upper block to slowly descend on to the end piece. A number of such frames are packed together in a carbon box, which is itself placed in a crucible. The space between box and crucible is filled with powdered carbon.

In the carbonization of amyloid threads, it has been found that in order to obtain the best results, the heat must be very gradually applied. Should the crucibles be exposed to a sudden increase of temperature, the filament contracting too suddenly, may be broken. Moreover, the gases produced by distillation might possibly deposit sufficient carbon on the sides of the filaments, to interfere with their homogeneity and also to cause them to cohere. Consequently, a pyrometer is frequently used in connection with the furnace in order to regulate its temperature. The time required to effect the carbonization will vary with the size and character of the filaments. Ordinary thread filaments may require eighteen hours for their proper carbonization, although a longer time may be employed. The period required for carbonization, necessarily varies with the size of the filaments, their character and the nature of the furnace.

PREPARATION OF THE FILAMENT.

Incandescent lamp filaments are now sometimes manufactured by a squirting process. This is in the main a secret process, but it consists essentially in obtaining an exceedingly intimate mixture of carbonaceous materials, forming them into a plastic mass and subjecting the same, while in the plastic condition, to powerful pressure, whereby they are forced or squirted through molds in die plates. Means are devised for preserving the shape which it is desired that this material should take, and then, after carefully drying the same, it is submitted to a suitable carbonizing process. Experience has shown that squirted filaments are capable of being rendered perfectly uniform. In all cases after the filaments have received the proper carbonization, it is necessary to wait until the furnace and its contents are cooled, before removing them from the carbonizing boxes, both on account of their fragile character, and for the sake of the furnace, which is apt to be injured by the sudden chilling.

CHAPTER VI.

MOUNTING AND TREATMENT OF FILAMENTS.

HAVING traced the filaments up to the completion of their carbonization, we will assume that they are ready to be removed from the carbonizing box. As already stated, the box and its contents are supposed to have cooled down to the temperature of the room. Care must be exercised in removing the filaments from the box, since the carbonization has changed the physical nature of the material, and the carbons are now brittle, though hard and elastic.

The next step in the manufacture of the lamp now begins viz., the *mounting* of the filament, or placing it upon a glass support through which the *leading-in wires*, or the conductors which carry the current to the filament are sealed. The ends of the filament are attached to the extremities of the leading-in wires by any suitable means. Much ingenuity has been displayed in obtaining a suitable mounting for the filament, and for a long time this mounting constituted the weak point in the lamp. Any collection of incandescent lamps, embracing specimens from an early period in the history of the art, would show how marked the evolution has been in this particular.

In order to obtain an idea of the relation of the various parts of the mounted filaments reference may be had to Fig. 12, which shows one of the common forms of mounting. A glass tube T, has a shoulder blown on it at S. Two copper wires w_1, w_2, ordinarily 0.015″ in diameter, are welded to short pieces of platinum wire,

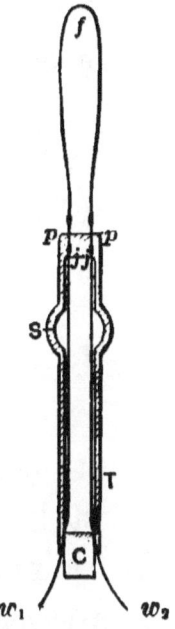

Fig. 12.—The Mounted Filament.

the method generally adopted being to hold the end of the copper wire against the end of the platinum wire in a flame, when they fuse together. These two wires constitute the leading-in wires. These wires are then laid in the glass tube *T*, and the glass is fused around the platinum wires in a flat seal at the point *S*, so that short projections of the platinum *p*, *p*, extend through the glass seal. The glass seal is then carefully annealed. The filament *f*, is now connected with its ends to the wires *p*, *p*.

When it is remembered that the operation of an electric lamp necessarily brings the filament to a white heat, it will be evident that means must be provided either for preventing the joints at *p*, *p*, from attaining a high temperature; or, if such temperature be attained, that the character of the joint must be such that it will not suffer. In any event it is clear that the seal *S*, must not be exposed to an incandescent temperature, and, therefore, the platinum extremities p, p, must remain comparatively cool. The method which is always adopted is to provide such a cross section at the joints p, p, that, taken in connection with the thermal conductivity of the platinum wires, the temperature of the joints will be much below the incandescing temperature of the rest of the filament. In the earlier forms of lamps, in which very large currents were usually employed, this result was accomplished by providing massive terminals connected with the carbon filament, possessing so great a radiating surface that their temperature was necessarily, comparatively low. Moreover, since most of these early forms of lamps did not employ a solid glass seal, it was still further necessary to reduce the temperature of the leading-in wires at the points of entry by means of cement.

Platinum is employed where the glass is fused around the leading-in wires, for

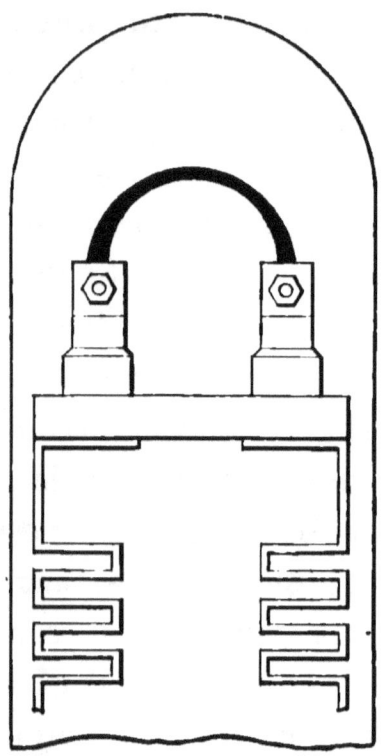

FIG. 13.—HORSESHOE LAMP.

the reason that the expansion and contraction of platinum is almost the same as that of glass. The expansion of glass rods varies from 0.0007 to 0.001 per cent. of their length, for each degree Centigrade, or just about the mean value of the expansion of platinum wires. When, therefore, glass is sealed around a platinum wire, and the seal is heated, the glass and the platinum expand together, and, on cooling contract together. If this were not the case, there would be a continual tendency to shear the glass over the platinum surface, and break the seal both on expansion and contraction. Among ordinary metals iron comes next to platinum as regards its expansion, and iron has been employed to some extent for the seal in incandescent lamps.

Fig. 13, represents an early form of lamp in which this method of connection was employed. A filament is connected to large terminals of copper supported on an insulating bridge. These terminals are further connected with long zig-zag strips of copper, extending to the base of the lamp, for the purpose of ensuring a low temperature where they pass through the base.

In some of the earlier lamps of the modern type the ends of the filaments were made of enlarged cross section, so that they were not only more readily secured to the platinum leading-in wires, but also by their increased mass prevented the current from raising it to the temperature of incandescence. In Fig. 14, a filament of bamboo is provided with enlarged ends. In some cases the carbon filaments, after they had been subjected to the carbonizing process, and while still in place in

TREATMENT OF FILAMENTS. 107

the carbonizing box, had their ends thickened by deposits of carbon, produced from

Fig. 14.—Bamboo Filament.

the decomposition of a carbonizable gas, forced into the mold at a certain stage in the process.

A very early form of joint consisted of a platinum bolt and nut, as shown in Fig. 13. A small screw lamp was sometimes employed for the same purpose as shown in Fig. 15. At one time, small metal blocks, *C, C*, were employed fastened upon the enlarged extremities of the filament, as shown in Fig. 16. Sometimes a small socket was formed in the end of the wire, and the carbon was placed in this receptacle and the socket pressed around it. In another form, the socket joint was secured more firmly to the carbon by covering it with an electrolytic coating of carbon, or the wire was wrapped around the carbon and subsequently secured to it in the same manner.

Fig. 15.—Early Form of Joint Lamp.

110   ELECTRIC INCANDESCENT LIGHTING.

This proved an excellent form of joint, and was employed extensively for a num-

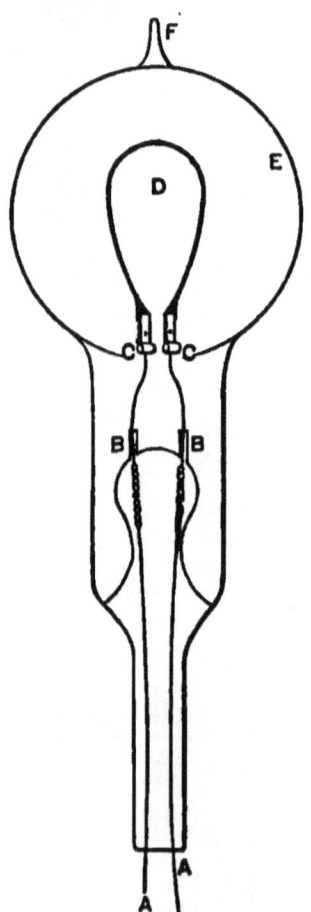

FIG. 16.—EARLY FORM OF CLAMP.

TREATMENT OF FILAMENTS

ber of years. It has, however, been replaced by a still simpler form in which no socket is employed, but one end of the filament is abutted against the end of the platinum wire, and carbon is deposited around the joint. This joint is obtained by dipping the abutting end into a suitable carbonizable liquid and sending a powerful current through the abutment while in the liquid. Under these conditions, a decomposition of the liquid occurs and hard carbon is deposited on the joint, thus effecting a thorough seal. At the present time the still simpler method is usually adopted of cementing the two together by a lump of carbon paste or dough.

In the early history of the modern incandescent lamp, the filaments, produced by substantially the processes already described, when placed in the exhausted lamp chamber and rendered incandescent by the passage of a current through them, were frequently found to glow irregularly, that is to say, there were parts which were rendered vividly incandescent while other parts were only dull red. This was due to the fact that the process did not produce homogeneous carbon filaments. In other words, either the diameter varied at different parts, or the resistivity of the material varied, or both. Consequently, when the incandescing current was sent through the filament, and the temperature gradually increased, the high resistance parts of the filament first began to glow, while the others remained comparatively cool, the heat being developed by the current in proportion to the resistance encountered. If such a spotted filament were employed in the lamp and the temperature raised by increasing the current, so as to make all parts of the filament glow, then the temperature of the high resistance portions would probably be raised beyond the limit of safety and the life of the lamp would consequently be much shortened; while, on the contrary, if the higher resistance portions of the filament were limited to the safe temperature, the lower resistance portions would be at so low a temperature, that the candle-power of the entire lamp would be unduly low.

This difficulty was happily overcome by an exceedingly ingenious process, generally called the *flashing process*, which consisted essentially in a means whereby carbon was caused to be deposited only upon those portions of the filaments whose resistance was higher than the rest. In other words, if the filament were unduly narrow at some particular spot, or had an undue resistivity at such spot, then carbon would be deposited upon this spot only.

The flashing process is carried on substantially as follows. The mounted filament is placed in a suitable chamber from which the air has been removed, and which is subsequently filled with a hydrocarbon vapor. An electric current, whose strength is gradually increased, is then sent through the filament. A hydrocarbon gas or vapor, suffers decomposition in the presence of a heated surface as soon as a certain temperature is reached. As the current gradually increases, the carbon filament begins to glow at its point of greatest resistance, and this point, consequently, receives a deposit of carbon, thus decreasing the resistance locally. If this current strength were maintained, the carbon would cease to glow at these points. If, however, the current strength be further increased, the carbon would begin to glow at the point of next highest resistance, and this in turn, receiving a deposit, would cease to glow at this current strength. It will be readily seen, therefore, that as the current strength is gradually increased, the filament receives a deposit at those portions of its mass only where it needs increase in conducting power, and soon the entire filament will glow with a uniform intensity of light.

It must not be supposed, that the carbon is now absolutely of the same area of cross section, or of the same thickness throughout. The flashing process has rendered it electrically, but not mechanically, homogeneous. We have correctly described the process as consisting of successive steps reached by gradually increasing the current strength. In point of fact, these steps follow one another so rapidly that the process at first sight may seem to be almost instantaneous, only a few seconds being required for an exceedingly spotted carbon to emit a uniform glow.

Although at the present day improvements in manufacture have resulted in the production of filaments, which are so nearly uniform in their resistance that they will glow uniformly when placed in the lamp, and, therefore, do not require to be subjected to the flashing process, nevertheless, since this process results in giving to the filament other valuable properties, it is still generally practiced. Not only are the surfaces of flashed carbon filaments harder than those which have not

TREATMENT OF FILAMENTS.   117

undergone this process, but the amount of light which they emit for a given current strength is markedly increased.

The flashing process is sometimes carried on in liquids, such as benzine, the filaments being dipped in the liquid and the current, as before, supplied in gradually increasing strength. In such cases, however, the decomposition of the liquid produces an atmosphere of gas around the filament so that the difference in the process is rather in appearance than in reality.

CHAPTER VII.

SEALING-IN AND EXHAUSTION.

THE mounted and flashed filament has now to be inserted in an enclosing glass chamber, in which it is hermetically sealed. This *sealing-in* is preferably accomplished by the actual fusion of the glass stem to the lower part of the globe. It will be interesting, therefore, to examine in detail the method generally employed in the manufacture of the incandescent lamp chamber and its hermetical closure on sealing-in.

Fig. 17, represents the successive steps that are generally taken in the sealing-in

SEALING-IN AND EXHAUSTION. 119

of the mounted filament in the lamp chamber. The glass lamp chamber A,

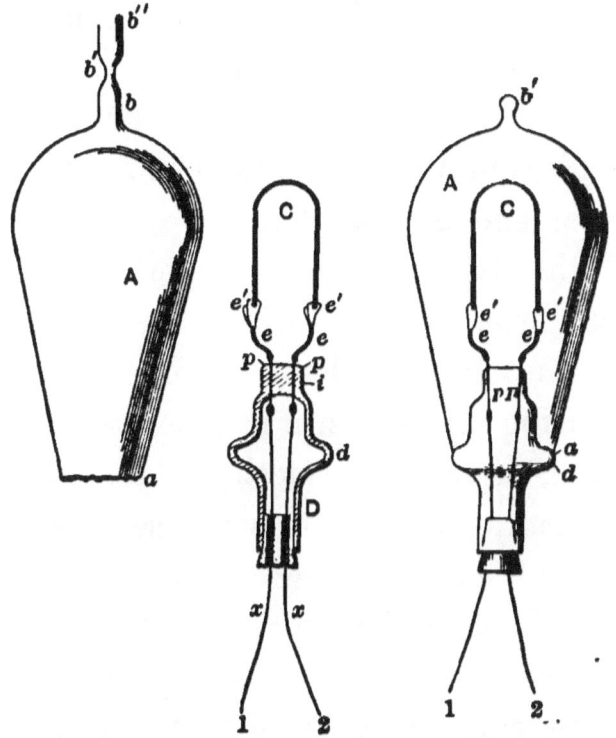

FIG. 17.—STEPS OF SEALING-IN PROCESS.

has the form shown, the open tubular projection being left at b, for the exhaustion of the chamber. The open end of the chamber A, is of such dimensions that the mounted filament can be introduced into it up to the shoulder d, which then rests in contact with the lower end a, of the chamber. The stem is then grasped by the glass-blower in one hand, and the tubular end of the chamber in the other, and the two revolved together as one piece, in a suitable blow-pipe flame directed upon the shoulder or joint, until the fusing temperature is reached, and the edge a, becomes hermetically sealed with the shoulder d. By this means it will be seen that an enclosing chamber, made entirely of glass, is provided with leading-in wires passing through the support at p, p. In the early history of the art, it was necessary that this delicate operation should be performed by a skilled glass-blower, but during recent years, machines have been introduced which grasp the

SEALING-IN AND EXHAUSTION.

globe and stem and revolve them in the blow-pipe flame with the requisite amount of pressure. The *machine seal*, so effected, is made as swiftly and neatly as that of the most expert workman. The sealed-in lamp is then carefully annealed, by subjecting it to the action of a gradually diminished heat, while under the action of a roller. Great care is necessary that the annealing of the joint should be thoroughly effected.

Attempts have been made, at different times, to produce a lamp in which a *mechanical seal* was effected between the stem and the globe, instead of a seal by fusion. Such a seal possesses the advantage of permitting the lamp to be readily repaired on the breaking of the filament. Figs. 18, 19, and 20, show a form of such *stopper-lamp*, as it is generally called.

122   ELECTRIC INCANDESCENT LIGHTING.

Fig. 18, represents the mounted filament, which is connected to the extremities of two iron wires sealed into the neck or

FIG. 18.—STOPPER-MOUNTED FILAMENT OF INCANDESCENT LAMP.

stem L. We have already pointed out that iron is sometimes employed for this purpose. The stopper portion of the stem

SEALING-IN AND EXHAUSTION.

at S, is ground by machinery to fit a similarly ground seat in the opening of the lamp globe, which is shown at A, Fig. 19. The mounted filament is inserted into the lamp chamber, and the stopper secured in its seat by a flexible cement. A brass shell is then secured around the base B of the lamp, enabling connections to be maintained with its terminals, as shown in Fig. 20.

The mounted filament, having thus been introduced into the lamp chamber, and the base of the lamp hermetically sealed, the next step is the exhaustion of the lamp chamber. An exceedingly small quantity of air left in the chamber will contain sufficient oxygen to cause rapid destruction of the lamp filament. Consequently, it is necessary to remove, as far as possible, all traces of air from the interior. This is

FIG. 19.—GLOBE OF STOPPER LAMP.

accomplished by means of pumps. The ordinary *mechanical air-pump*, of the automatic valve type; *i. e.*, in which the valves are opened and closed automatically by the motion of the piston, is capable of producing fairly high vacua, but not sufficiently high for the purposes of being used alone in the exhaustion of the lamps, since the residual air would still be detrimental. Mechanical pumps, however, are often used for producing a rapid exhaustion of the lamp chamber, the final exhaustion being accomplished by means of a *mercury pump*. The operation of the mercury pump, however, is so simple in practice and efficient in action, that in some cases, the use of the mechanical pump is dispensed with, and the entire process of exhaustion is carried on by the mercury pump alone.

A great variety of mercury pumps have

Fig. 20.—Completed Stopper Lamp.

been devised, all of which may be conveniently divided into two classes; namely, those of the Geissler type, in which the vacuum is obtained by utilizing the principle of the so-called Torricellian vacuum, of the barometer tube, and those of the Sprengel type, in which the vacuum is obtained by the fall of a stream of mercury. When a column of mercury is permitted to fall through a vertical tube, connected near its upper end by a branch tube with the chamber to be exhausted, the air will be carried away from the chamber by becoming entangled as bubbles in the falling column. Mercury pumps of this character are well adapted to the exhaustion of lamps, owing to the simplicity and efficiency of their action.

When the proper degree of vacuum is obtained, the lamp is *sealed-off* by fusing the tube at the top of the lamp chamber, with a blow-pipe flame, a constriction being provided in the tubulure b, at b^1 as shown in Fig. 17, for facilitating this process. In the early state of the art this sealing-off was effected while both the lamp and the filament were cold. It was found, when thus sealed-off, that, although the proper vacuum had been obtained, and the lamp operated satisfactorily for a while, yet the vacuum soon invariably deteriorated, so that the life of the lamp was unduly shortened. The explanation was at last found in the fact that gases were occluded or absorbed in the carbon filament, as well as condensed on the inner surface of the globe. These gases adhered to the filament, or to the globe, with too great a force to permit them to escape into the lamp chamber during the process of exhaustion. When, however, the lamp was lighted after concluding the process of exhaustion, the intense heat of the filament disengaged these gases which reduced the vacuum injuriously. The remedy is simple. As soon as a fairly good vacuum is obtained in the lamp, an electric current is sent through the filament and the last stages of the pumping process are carried on while the filament is aglow. Immediately before sealing-off, the current is increased beyond the strength intended to be employed in practice, and the lamps sealed while the pumping is carried on. By this means the occluded gases in the filament and on the globe are carried off, and this process has done much to improve the lamp.

When incandescent lamps were first manufactured on a large scale, an effort was made to obtain as high a. vacuum as possible, and pumps were employed which, it was claimed, left a residual atmosphere in the lamp chamber of but 1,000,000th of its original amount; that is that 999,999 parts of air out of every million had been removed. Even at the present time, while it is generally considered that residual atmospheres of air, containing as they necessarily must traces of oxygen, result in a decreased life of the lamp, yet it may be asserted, as a result of actual experience, that residual atmospheres of certain gases, such as chlorine or bromine, or mixtures of the same, may not only be innocuous but actually advantageous.

It is quite possible, during the operation of a lamp, the sealing-off of which has been thoroughly made, that the vacuum may actually improve rather than deteriorate; for, in the disintegration of the carbon, to which we shall allude in a subsequent chapter, a deposit of finely divided carbon takes place inside the lamp chamber. This carbon possessing, as it does, the power of occluding or absorbing residual atmospheres, would naturally tend to improve the vacuum during use.

The exhausted and sealed lamp must now be provided with a base, whereby it can be readily placed in a socket or support. The lamp base is so arranged that two metallic portions, suitably insulated from one another, are connected to the ends of the leading-in wires. These portions on the base are adapted to connect the lamp with the circuit wires by the mere act of placing it in the lamp socket. Various forms of lamp bases are employed as we shall see hereafter. The lamp base is attached to the lamp by a cement, generally of plaster of Paris.

132 ELECTRIC INCANDESCENT LIGHTING.

Some of the more usual forms of lamp bases are shown in Fig. 21 at *A*, *B*, *C* and *D*. The two separate metallic pieces, which are electrically connected to the ends of the leading-in wires, are, in all

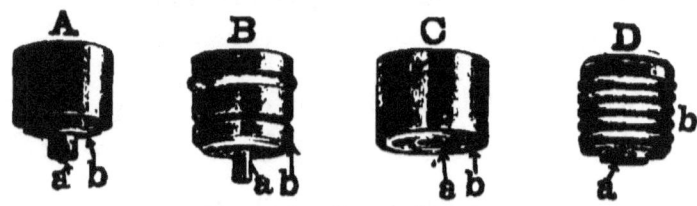

FIG. 21.—LAMP BASES.

cases, indicated by the letters *a* and *b*. An inspection of the figure will show that *A*, has a central contact pin as one terminal, and a concentric brass ring as the other terminal. *B*, has a central pin as one terminal, and an external concentric cylinder as the other. At *C*, two concentric cylinders are employed, the inner one has, however, a screw thread for securing it in its socket. *D*, has a screw shell as one terminal, and a central cap as the other terminal.

It is sometimes convenient to fit a lamp of one manufacture into a socket of

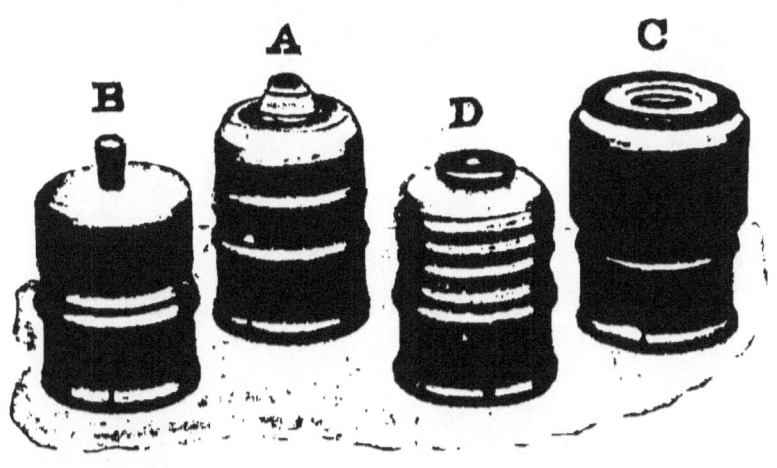

Fig. 22.—Lamp Adapters.

another manufacture. For this purpose a device called a *lamp adapter* is employed. An adapter consists essentially of an exterior base and an interior connection piece.

In Fig. 22, four adapters are shown suitable for attachment to a stopper lamp, and provided with bases corresponding to the lamp bases shown in *A, B, C, D,* of Fig. 21, similar letters corresponding in the two figures. The use of stopper lamps has, however, been abandoned.

CHAPTER VIII.

LAMP FITTINGS.

On the removal of the lamp from the pumps, it is tested for candle-power, and marked in volts for the pressure which should be supplied to it in operation.

Some examples of finished lamps are shown in Fig. 23. These lamps are all of the same type and differ only in the form of base. The bases of these lamps differ so as to permit each lamp to be used on some particular socket.

One of the simplest forms of sockets intended for inexpensive work, especially

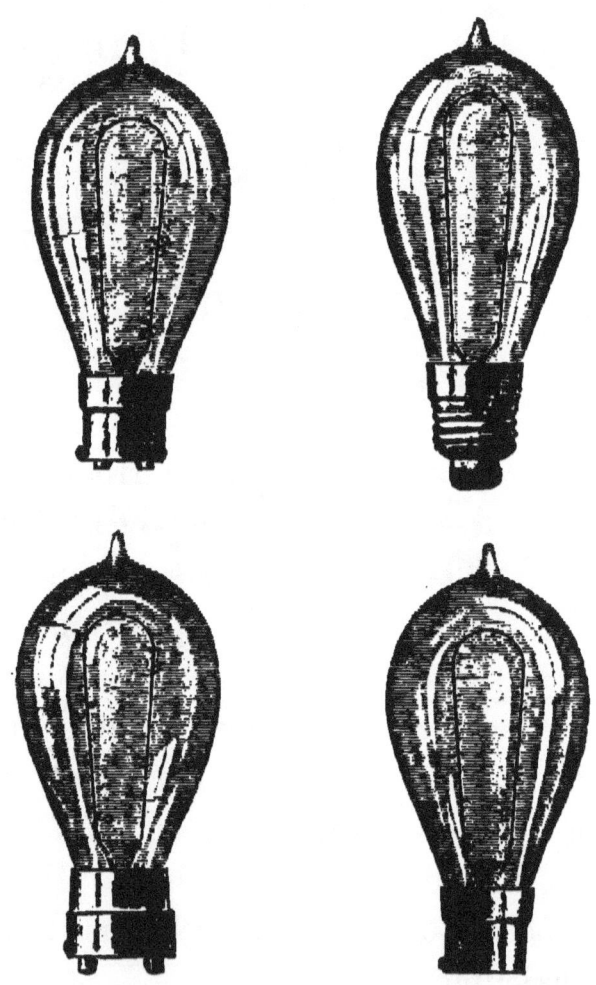

Fig. 23.—Completed Incandescent Lamps.

where the lamps are not open to direct observation, such as at the foot or side lights of theatres, is shown in Fig. 24. This

Fig. 24.—Simple Form of Socket.

socket is intended to receive a lamp with a screw base. A wooden shell L, is screwed to the wall or other support, by ordinary screws passing through the screw holes, one of which is seen at S. A and B, are brass screws intended for the reception of the circuit wires or mains, and are connected respectively to the brass screw shell

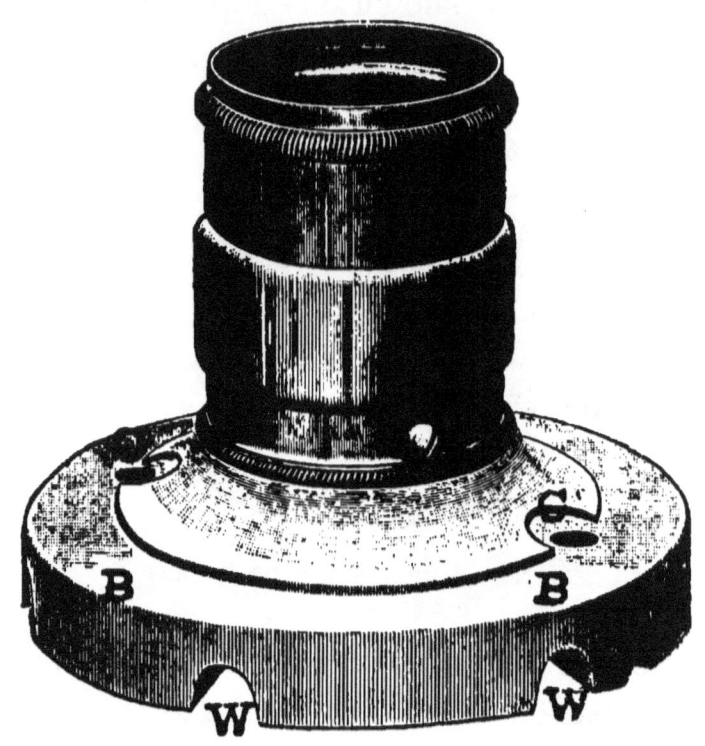

Fig. 25.—Keyless Wall-Socket.

C, and a central cap beneath. These will make contact with the two insulated portions of the base of the lamp when the lamp is screwed in.

FIG. 26.—KEY WALL-SOCKET.

Fig. 25, shows a more sightly form of *keyless socket*, that is, a socket which is not provided with a key or switch, and which, therefore, constantly maintains connection between the mains and the lamp. The base *B B*, is of porcelain, and is fixed in position by screws passing through screw holes *C, C.* Supply wires, connected with the mains, pass beneath through the grooves *W, W.* Connections are maintained through the interior of the shell.

A socket of a similar character, but provided with a key *K*, is shown in Fig. 26. In this case the lamp is lighted or extinguished by the turning of the key. In the case of keyless sockets the lamps are turned on or off by the action of a distant *switch.*

Socket keys, open and close the circuit of a lamp at one point in a variety of ways. Two forms of socket keys are shown in

LAMP FITTINGS. 141

Figs. 27 and 28. Fig. 27, shows a socket suitable for use with a lamp base of type *B*, Fig. 21; and Fig. 28, shows a socket suitable for use with the lamp base of type

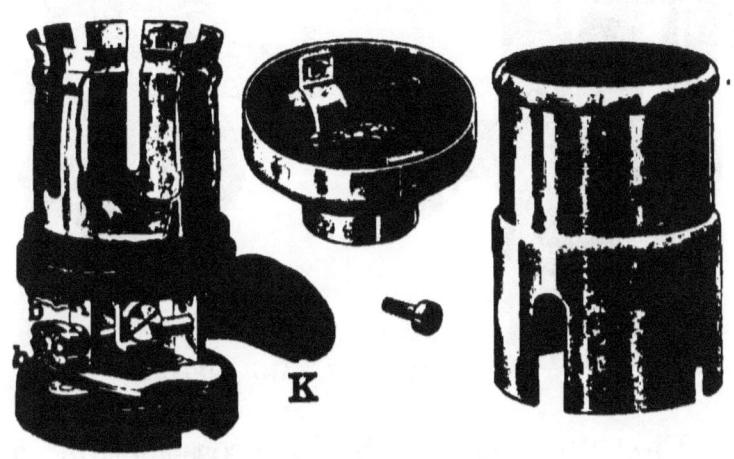

FIG. 27.—DETAILS OF SOCKET.

C, in that figure. In Fig. 27, the turning of the key *K*, makes contact between the brass segments *b, b*, through the intermediary of the cam, on the extremity of the key axis. In Fig. 28, a similar method is

142   ELECTRIC INCANDESCENT LIGHTING.

employed. Here the movement of the key closes connection between contacts *b, b,* through the metal piece *C.*

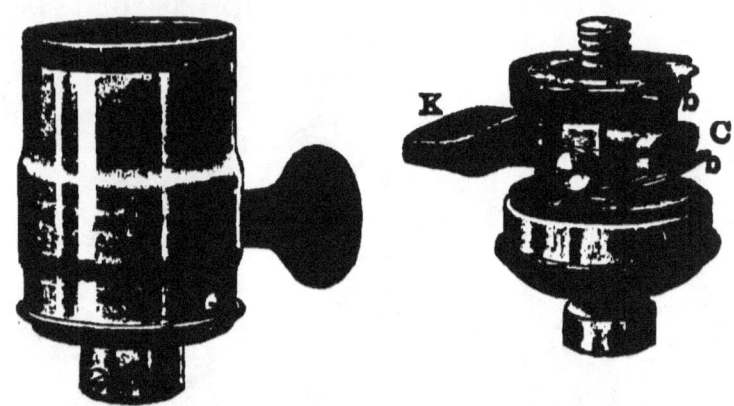

Fig. 28.—Details of Lamp Socket.

Fig. 29, represents in cross-section a lamp with a screw base in its socket. The current is turned on or off at the key *K. G,* is the globe, and *T,* the tip at which the lamp was sealed on removal from the pump; *F,* is the filament; *S,* the seal of the leading-in wires; *W,* the welds between the platinum and the copper lead-

LAMP FITTINGS.

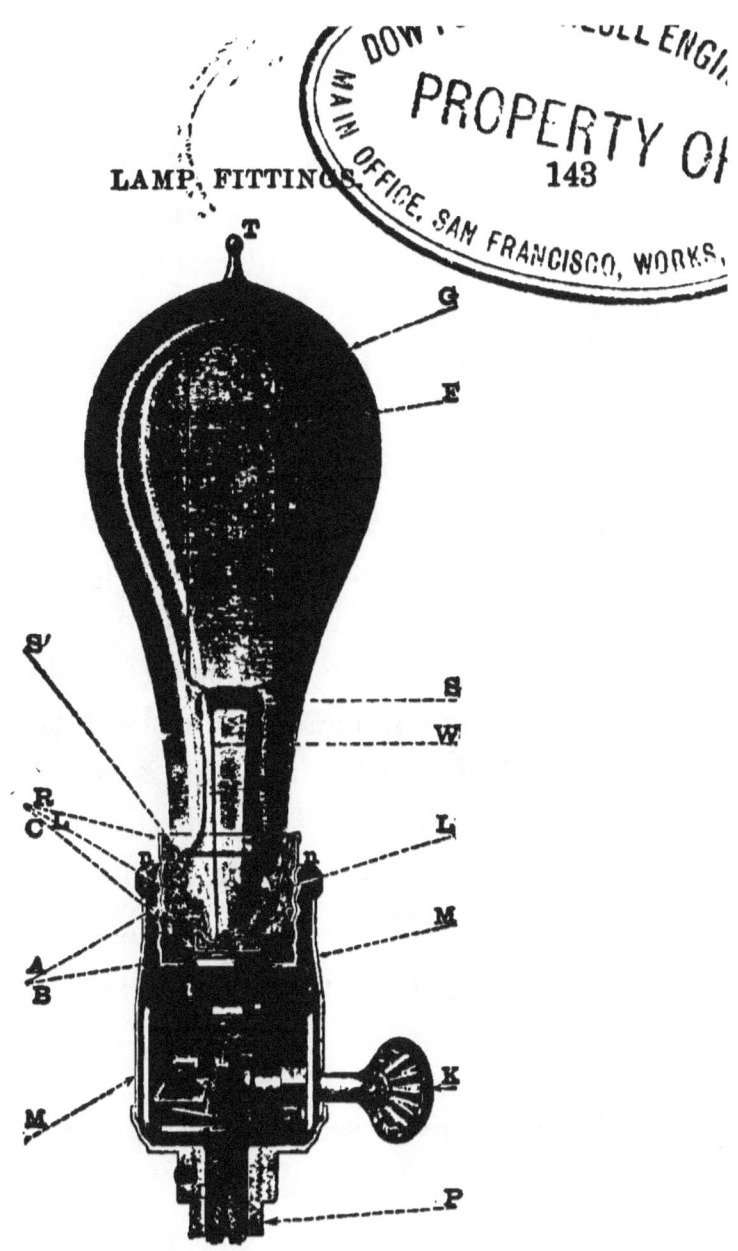

FIG. 29.—LAMP AND SOCKET SHOWING CONNECTIONS.

ing-in wires; S', the seal of the lamp stem with the glass chamber. $L, L,$ are projections of the glass on the surface of the shoulder, intended to aid in securing the lamp in its plaster of Paris cement. C and $R,$ are the cap and screw metallic pieces of the base, each in soldered connection with one leading-in wire as shown. A and $B,$ are the cap and screw connections of the socket, each in connection with one of the external wires entering the socket through the pipe or metallic support $P.$ One of these wires is connected directly with the brass shell $A,$ while the other is connected with the cap $B,$ only through the intermediary of the key $K.$ $MM,$ is the external brass shell of the socket, insulated from the interior portions by the hard rubber ring $n\ n.$

When an incandescent lamp is supported

LAMP FITTINGS. 145

by a flexible cord, it usually requires both hands to turn the key at the socket on or off, one to hold the lamp, and the other to

FIG. 30.—PUSH BUTTON KEY SOCKET.

turn the key. Fig. 30, shows a device by which the turning on or off can be accomplished by the hand which holds the socket. This is accomplished in replacing the ordinary key by two pressure buttons. Pressure upon the stud *A*, forces it home until it is locked by the trigger *B*, thus

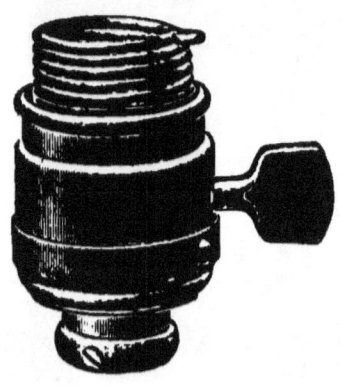

FIG. 31.—SPRING SOCKET FOR SCREW BASE.

turning on the lamp and keeping it turned on. Pressure on the trigger *B*, releases the push *A*, which returns to its original position under the action of a spring.

Fig. 31, shows a form of spring socket for a screw lamp base. For temporary work,

LAMP FITTINGS. 147

such as exhibitions, etc., where the expense of permanent fixtures would not be justified, a temporary socket is sometimes employed, as shown in Fig. 32. It consists of

FIG. 32.—TEMPORARY SOCKETS.

a spiral spring holding the screw base of the lamp, and connected with one supply wire W, while the brass screw in the centre of the spiral is connected with a second supply wire beneath the wooden base bar

Such construction is, of course, not regarded as safe for permanent use in buildings.

Where lamps are placed in positions exposed to the weather, it is necessary to

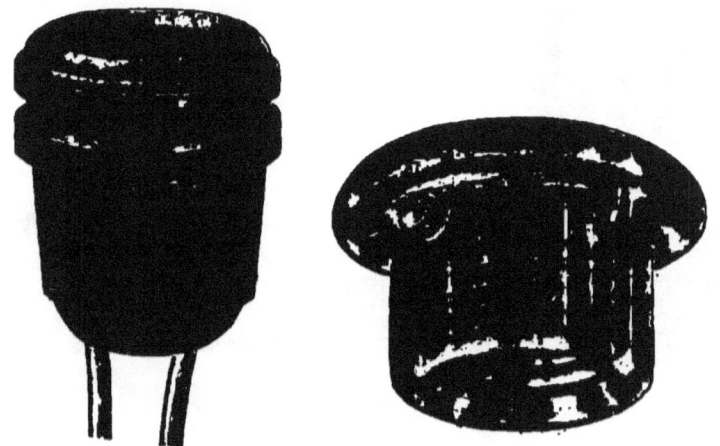

FIG. 33.—WEATHER-PROOF SOCKETS.

employ some form of weather-proof socket. Two forms of such sockets in common use are shown in Fig. 33, that on the right hand side being of glass, and that on the left,

LAMP FITTINGS. 149

of hard rubber or composition. These sockets are intended for the reception of screw bases.

Various forms of reflecting surfaces are employed in connection with the lamps so

FIG. 34.—METALLIC SHADE FOR REFLECTING LIGHT DOWNWARDS.

as to throw the light in any desired direction. Two such forms of *lamp shades*, devised to throw the light downwards, are shown in Figs. 34 and 35. The depth of the shade will vary with the character of the illumination required. That shown in Fig. 34, is suitable for the illumination of desks from above. Fig. 35, is suitable

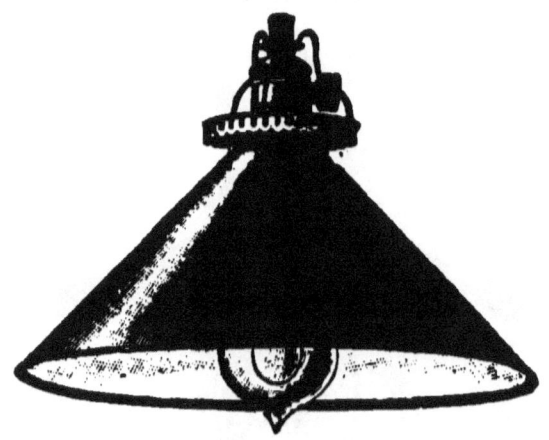

FIG. 35.—METALLIC SHADE FOR REFLECTING LIGHT DOWNWARDS.

for the illumination of a billiard table. Fig. 36, shows a form of *half shade* sometimes employed for desk use. Here one half of the lamp only is covered by the shade which in outline conforms with the shape of the lamp, the inside of the shade being provided with a good reflecting surface to throw the light downwards.

Fig. 36.—Metal Half Shade for Desk Use.

It is important in order to ensure good illumination for reading, that all portions of the printed page shall be equally lighted. Although this is generally secured by the aid of any good shade, yet it is often preferable to employ for this purpose the form of shade and enclosing globe shown in Fig. 37.

Fig. 37.—Reflector Shade.

Instead of forming the lamp shade out of a single surface, a number of surfaces are sometimes employed, either plane or curved. Figs. 38 and 39 represent

LAMP FITTINGS. 153

panel reflectors; i. e., reflectors composed of strips or panels of silvered glass. These

FIG. 38.—CONCAVE PANEL SHADE AND REFLECTOR.

reflectors are suitable for store windows or railway stations. The shade in Fig. 38, is designed to throw light downwards from

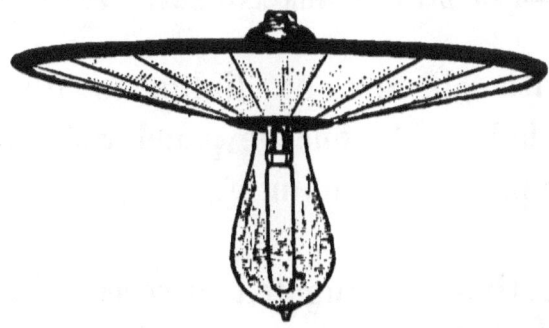

FIG. 39.—CONCAVE PANEL SHADE AND REFLECTOR.

its concave surface, and Fig. 39 to throw the light outwards from its convex surface as well as downwards. Finally, reflectors of similar forms are also arranged to oper-

Fig. 40.—Panel Reflector and Shade for Cluster.

ate with clusters of lights to illumine larger halls. A concave panel reflector of this type is shown in Fig. 40.

Sometimes corrugated silvered glass is employed in various forms for reflecting

LAMP FITTINGS. 155

purposes. Fig 41, shows a reflector of this type made in the form of a shade.

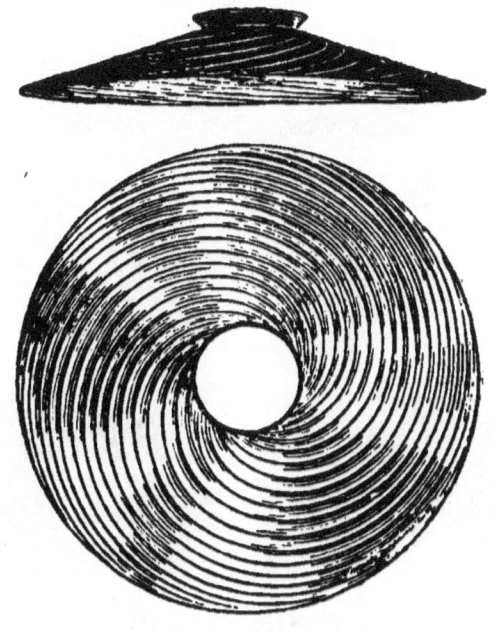

FIG. 41.—CORRUGATED REFLECTOR AND SHADE.

Even transparent or translucent substances may at times be employed for reflectors. In such cases they aid in scattering light downwards as well as in reflecting it. Materials employed for this purpose are glass and porcelain, either plain or corrugated, transparent or trans-

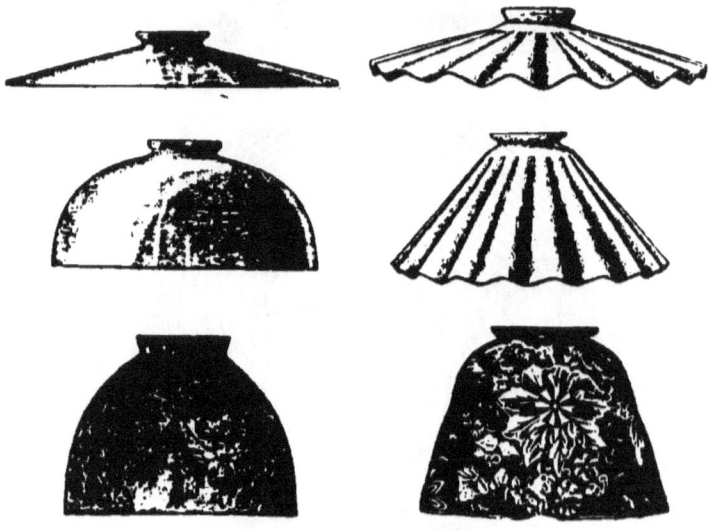

FIG. 42.—GLASS SHADES.

lucent. Fig. 42, shows a variety of shades employed for such purposes.

Owing to the fragile nature of the incandescent lamp chamber, it is necessary,

LAMP FITTINGS. 157

when lamps are placed in exposed positions, to protect them from accidental destruction by blows. For this purpose *wire gratings* or *shields* are arranged, of

FIG. 43.—HALF WIRE-GUARD.

such form and outline as shall not seriously interfere with the light. Such gratings would, of course, be inadmissible in any situation where marked shadows are objectionable, and should, therefore, only be employed in cellars, underground passages, or in similar situations. Fig. 43, shows a form of *half wire-guard*, which,

FIG. 44.—FULL WIRE-GUARD.

as its name indicates, only surrounds a portion of the lamp proper. Fig. 44, shows a full wire-guard surrounding the entire lamp. In Figs. 45 and 46, some

LAMP FITTINGS.   159

forms of wire-guards are shown, suitable for portable or hanging lamps.

Sometimes, instead of employing a wire guard to protect the lamp, the lamp is

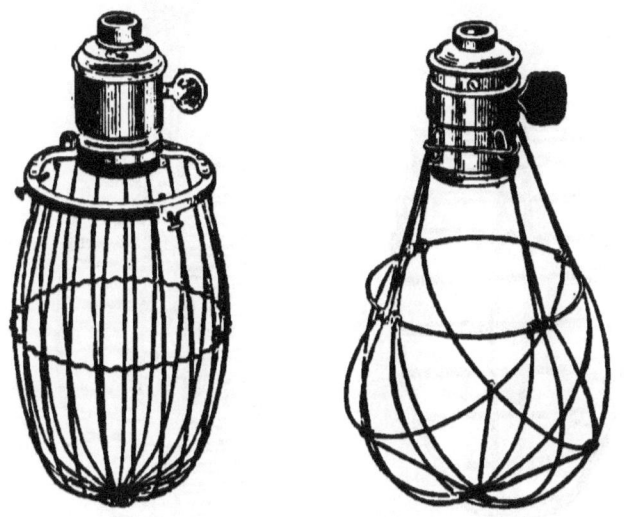

Fig. 45.—Wire-Guards for Suspended Lamps.

entirely surrounded by an air-tight or a steam-tight glass lamp chamber as in Fig. 47. This globe consists of two parts, a cylindrical glass cover with a rounded

160 ELECTRIC INCANDESCENT LIGHTING.

end, provided with a screw thread metallic cap, capable of tightly fitting into the base

FIG. 46.—PORTABLE LAMP GUARD.

as shown. A steam-tight lamp chamber is employed under circumstances where it is either desired to protect the lamp from cor-

LAMP FITTINGS. 161

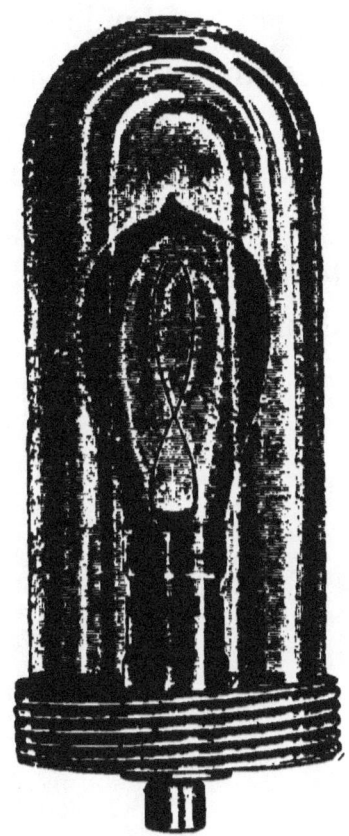

Fig. 47.—Steam-Tight Lamp Chamber.

rosive vapors, from spray at sea, or for the purpose of avoiding any possible accident which might result, from the explosion of inflammable gases or clouds of dust, on the accidental breaking of the lamp chamber and globe. It will, of course, be understood that the use of any form of wire grating, or steam-tight globe, must necessarily be attended by a marked decrease in the useful illumination of the lamp.

CHAPTER IX.

THE INCANDESCING LAMP.

WE have now traced the manufacture of the lamp up to the time when it is ready to be connected with the supply wires and actuated by the electric current. With this the manufacture of the incandescent lamp, that is a lamp capable of being rendered incandescent, now ceases, and the story of the *incandescing lamp*, and the theory of its operation begins. We will, therefore, trace in this chapter such theoretical considerations as enter into the action of the incandescing lamp.

An incandescing lamp may be regarded as an electro-receptive device, wherein the energy of the electric current is being converted into heat energy, part of which is luminous. It is, therefore, necessary, at the outset, to determine the amount of energy which the lamp is absorbing. This, as we have already seen, is equal to the product of the number of amperes, or the current strength supplied to the lamp, and the pressure in volts at the lamp terminals. If the resistance of a lamp, when hot, is 200 ohms, and the pressure between the mains 100 volts, then the current which will pass through the lamp will be $\frac{100}{200}$ = 1/2 ampere, and the activity supplied will be 1/2 × 100 = 50 watts or $\frac{50}{746}$ th horse-power.

The temperature which an activity of 50 watts, will produce in a lamp filament, depends both upon the extent of the surface area of the filament and upon the character of its surface. If the surface of the filament be large, the activity per square inch, or per square centimetre, will be comparatively small, and the filament will not be highly heated. If, on the other hand, the surface area of the filament be small, the intensity of its surface activity will be great, and the temperature of the filament will be high. The surface activity of an incandescing filament is, approximately, from 70 to 100 watts per square centimetre, or 450 to 650 watts per square inch; *i. e.*, about $\frac{6}{10}$ths to $\frac{8}{10}$ths of a horse-power per square inch of surface.

The electric arc lamp has in its crater, a

surface activity of, approximately, 3 KW, or 3,000 watts per square centimetre of surface; *i. e.*, 19.35 KW per square inch, or 27.15 HP per square inch. This shows that the surface area of the crater is very small, since an arc lamp takes only about 3/5ths horse-power. The apparent surface activity of the sun is, approximately, 10 KW, or 13.4 HP per square centimetre; *i. e.*, 64.5 KW, or 86.5 HP per square inch. Since, as we have shown, the highest frequency of the waves which are emitted by a heated surface increases with the temperature, it is evident that the highest frequency reached in the solar light waves will be greater than in the arc light waves, and this in its turn will be greater than in the incandescent light waves, since the intensity of surface activity in watts per square centimetre differs so markedly in these three surfaces.

Collecting these results tabularly, we obtain,

Solar surface activity, . . . 10,000 watts per square cm.
Arc-crater surface activity, . . 3,000 watts per square cm.
Incandescing-filament surface activity, 70 to 100 watts per square cm.

If the surface activity of an incandescing filament be increased, by passing a stronger current through the filament, that is, by subjecting it to a higher electric pressure, the temperature of the filament will increase, and with it the amount of light given off per square centimetre. This increase may be carried up to the point of destruction of the carbon filament. The *duration* or *life* of an incandescing lamp, depends very markedly upon the temperature and surface activity of the filament. At a low temperature, or at a dull red heat, an incandescing lamp will last almost indefinitely; at a vivid incandescence, or very high temperature and intense surface activity, its life may be only a few minutes. Between these two extremes lies a mean temperature and surface activity, at which it has been found in practice most profitable and desirable to operate the lamp. The object, therefore, of the lamp maker is so to proportion the dimensions of the filament, that, when connected across the mains, the surface activity will reach the amount necessary to give the proper temperature to the filament, as well as the desired total quantity of light.

It should be carefully remembered that the surface activity, which determines the brightness of the filament, is a quantity altogether distinct from the *total-candle-power*, or the total quantity of light given off from the lamp. The *brilliancy* depends only on the surface activity, while the total candle-power depends upon the total surface area as well as upon the brilliancy. It is common to find that a person looking at two lamps, one of which may have a high surface activity; *i. e.*, a great brilliancy, but which only gives say 5-candle-power, and the other of which has a low surface activity, or small brilliancy, but which gives 16-candle-power, will judge that the brighter lamp is giving the greater amount of light. In other words, the eye is very sensitive to relative brightness or brilliancy, but is by no means sensitive to differences in *total-candle-power ; i. e.*, total intensity of the light. It is very essential, in all artistic groupings or arrangements of lamps, that their surface activity and brilliancy should be as nearly equal as possible, since, otherwise, appearing to the eye unequally bright, they will fail to produce pleasing effects.

Having given a certain temperature and surface activity to the filament, the total amount of light will depend upon the total surface area. For example, a 16-candle-power lamp, operated at a given temperature, or at an *efficiency* of say 1/3 candle-per-watt, would take 48 watts of activity. A 32-candle-power lamp, at the same brightness, surface activity, and quality of carbon filament, and therefore, with the same efficiency of 1/3 candle-power per watt, would require to be supplied with 96 watts and would, therefore, require double the surface from which to radiate the doubled activity. Such a lamp, working at the same pressure, would require to be both larger in diameter and longer. Broadly speaking, a lamp of high candle-power will have a thick filament, and a lamp of low candle-power a thin filament, when working at a common pres-

THE INCANDESCING LAMP. 171

sure. Fig. 48, represents the relative sizes of incandescent lamps of the same manufacture, voltage and efficiency, intended for 16, 32, and 100 candles. Here the gradually increasing lengths and diameters of the filaments may be observed.

It may be well to explain in greater detail the meaning of the term efficiency, as used in the preceding paragraph. Since energy must be expended in an incandescing lamp, in order to produce a certain candle-power, it is evident, from the standpoint of economy in energy, that the greater the number of candles which can be obtained per horse-power, or per watt of activity, the greater will be the efficiency of the lamp. We speak, therefore, of the efficiency of an incandescent lamp as being 1/3rd or 1/4th of a candle per-watt, meaning that an 8 candle-power

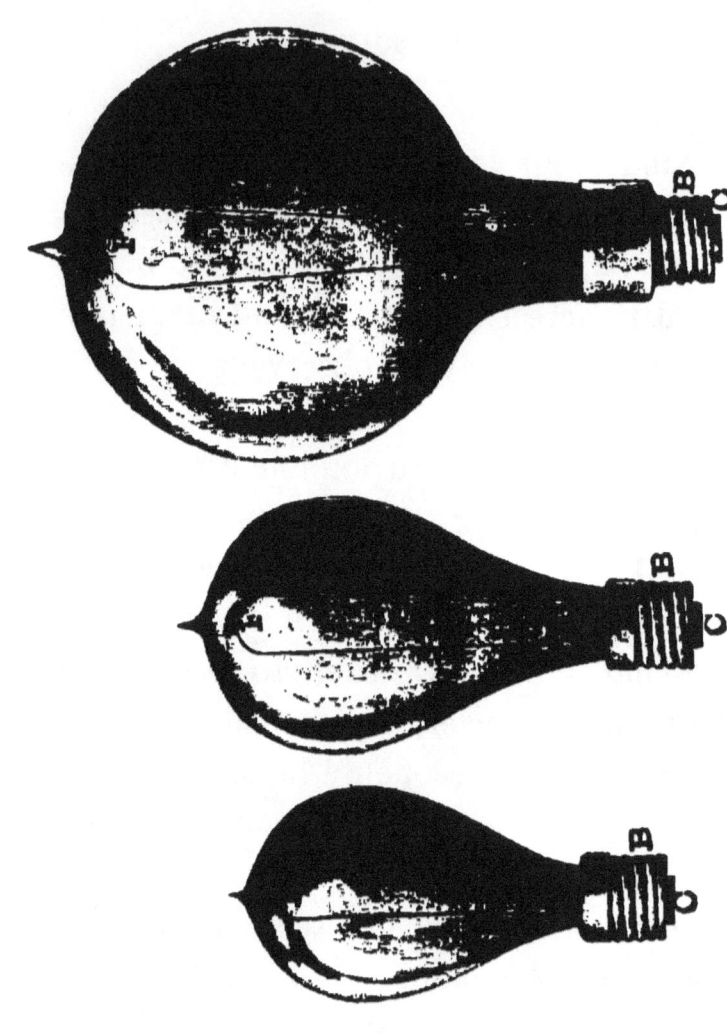

Fig. 48.—16-Candle-Power, 32-Candle-Power and 100-Candle-Power 100-Volt Lamps.

lamp would take, in either case, 24 or 32 watts respectively, representing 248.7 or 186.5 candles per electrical horse-power. In common usage, however, the term efficiency is often unfortunately misapplied, so that the same lamps would be spoken of as having an efficiency of 3 or 4 watts per candle, respectively, from which it would seem that as we increase the number of watts to the candle we increase the efficiency, whereas, it is evident that the reverse is true. It is preferable, therefore, to use the word efficiency in the less popular but more correct signification.

When an incandescent lamp is operated at a constant pressure, a series of changes takes place which it is important to follow as closely as the knowledge we possess will permit.

The temperature reached by the filament of an incandescing lamp has been estimated from a series of measurements, to be in the neighborhood of 1,350° C., slightly varying, however, with the surface activity and brightness.

Thus, at an efficiency of 1/3 candle per watt, the temperature is estimated to be 1,345° C.

At an efficiency of 1/4 candle per watt, the temperature is estimated to be 1,310° C.

And at an efficiency of 1/4.5 candle per watt, the temperature is estimated to be 1,290° C.

If we increase the activity of an incandescing lamp one per cent.; *i. e.*, if we increase the pressure at its terminals to such a point that the number of watts it receives increases by one per cent., the temperature is believed to increase about 2° C.

and the candle-power is believed to increase about three per cent.

When the electric current passing through a lamp produces the surface activity and temperature for which the lamp is designed, although, in general, the lamp exhibits a steady diminution in temperature, surface activity and candle-power, which continues while the lamp is used, yet it frequently happens, that for the first few hours these quantities actually increase, so that a 16-candle-power lamp, after the first fifty hours of its life, may give 17 candles, and a greater brightness than at the start. Even when this rise occurs, however, at the end of the first hundred hours the lamp will usually have fallen in candle-power, brilliancy, surface activity and temperature, all these quantities being associated, to an amount varying with the type of lamp and the carbon from which it has been manufactured.

We shall now examine the causes which bring about the progressive decay of the lamp above referred to. When an incandescent lamp is operated, it is found that the negative half of the filament throws off or projects carbon particles from the surface, in all directions in straight lines. It is generally believed that this effect is due to a species of evaporation. This evaporation takes place with greatest activity near the negative extremity of the filament, or the point where the filament is united with the leading-in wire on the negative side. From this point of maximum evaporation, the effect diminishes to the centre of the filament and the positive side of the filament shows but little evaporation. The presence of *electric evaporation* from negatively charged surfaces is recognized at all temperatures, and even under atmospheric pressures, but, like all evaporation, is aided by a high temperature and vacuum. Consequently, its effect is pronounced in an incandescent lamp.

The existence of an evaporation of the filament can be detected in a variety of ways. For example, if a metallic plate be supported in the lamp chamber, midway between the two legs of an ordinary horse-shoe filament, it is found that the evaporation of the carbon particles from the negative side of the filament, is actually capable of carrying an electric current to the plate. That is to say, the stream of negatively charged carbon particles impinging against the surface of the metallic plate, delivers up to it the electricity with which they are charged and so results in the passage of an electric current.

The continued evaporation of carbon from the surface of the filament, produces a gradually increasing blackening of the surface of the globe, since the projected carbon particles adhere to the walls of the chamber where they strike. The continued bombardment of the glass walls slowly coats them with a layer of carbon, which being opaque, reduces the amount of light emitted by the lamp. An old lamp, if examined against a white surface, such as a sheet of paper, will be seen to be distinctly blackened over its interior surface. This *blackening*, or, as it is sometimes called, *age-coating* of the lamp chamber, takes place, other things being equal, most rapidly in lamps that have been burned at a high temperature, since in these the evaporation is more rapid.

Another proof that the particles of carbon leave the surface in straight lines is to be found in the "*shadows*" produced on the surface of the glass, when the filaments are straight horse-shoes, or lie wholly in one plane. In such cases the bombardment from any portion of the negative leg is necessarily intercepted by the positive leg in the plane of the filament, so that the globe is protected in this plane by the interposition of the positive leg. The result is, that after the lamp chamber is visibly blackened, a distinctly marked line can be traced on the surface of the glass opposite the negative leg. The term shadow is sometimes applied to this. There will, however, be no shadow on the side of the glass nearest to the negative leg, nor will the shadow be produced if the polarity of the supply mains is occasionally reversed, or, in the case of lamps supplied by alternating currents where each leg is positive and negative alternately.

Not only does a lamp filament give less light after being operated at high pressures for a considerable length of time, owing to an increase in its resistance and the blackening of the globe, but also to a change in the surface nature of the filament, whereby it gives less light for a given surface activity. In technical language its *emissivity* increases, so that the temperature, which is attained by a given surface activity, is reduced. In other words, the lower the emissivity of a filament, the higher the candle-power and brilliancy for a given surface activity. Of these three

THE INCANDESCING LAMP. 181

causes for decreased candle-power with age; namely, diminished current strength and activity, diminished translucency of the globe, and increased emissivity, the loss in candle-power is about equally affected by each.

Fig. 49 shows curves representing the change in candle-power of an ordinary incandescent lamp when initially operated at various efficiencies. Thus, at an efficiency of 0.3 candle-per-watt, the lamp gives 14.5 candle-power. At an efficiency of 0.4 candle-per-watt, 22.5 candle-power, and at 0.5 candle-per-watt, 32 candle-power. Roughly speaking, if we double the efficiency we treble the candle-power and brilliancy of the lamp. On this account it would obviously be advantageous to increase the efficiency of a lamp as far as possible.

182 ELECTRIC INCANDESCENT LIGHTING.

In the preceding diagram we have plotted the relation between candle-power and efficiency. If, however, we plot the relation between candle-power and activity,

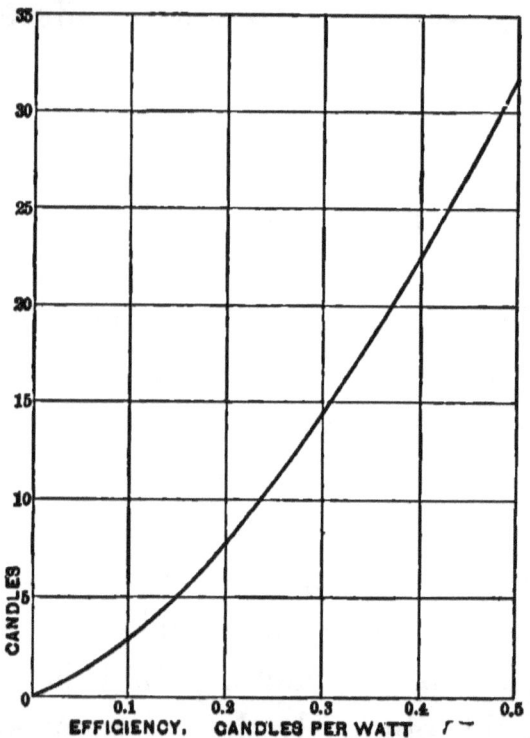

FIG. 49.—CURVE OF RELATIVE CANDLE-POWER OR BRIGHTNESS FOR A PARTICULAR CHARACTER OF CARBON FILAMENT OPERATED AT DIFFERENT EFFICIENCIES.

we find, roughly speaking, that the candle-power increases as the cube of the activity, so that if we double the activity of the lamp; *i. e.*, increase the pressure at its terminals until the product of this increase of pressure and the increased current is double what it was originally, the candle-power of the lamp will be increased about eight times, giving about four times as much light per horse-power or per watt expended in the lamp.

Fig. 50 shows the relation, which has been experimentally found to exist, between the average lifetime of lamps of the same make as that represented in Fig. 49, and the efficiency at which such lamps are burned. A study of this curve will show why it is not, at present, possible to obtain in practice an efficiency beyond a certain value; for, at an efficiency of 0.2 candle-per-watt, corresponding to 8 candles in the particular lamp of Fig. 49, the mean duration of life is over 16,000 hours; while at 0.5 candle-per-watt, corresponding to 32

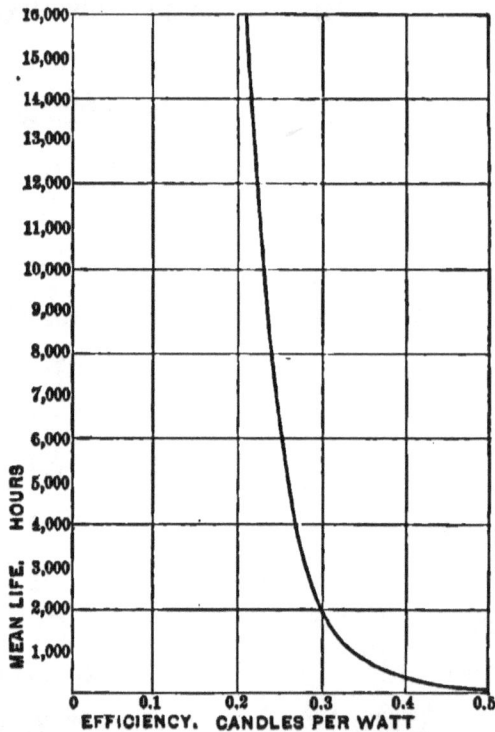

FIG. 50.—CURVE SHOWING THE MEAN DURATION OF LIFE IN A PARTICULAR CLASS OF INCANDESCING LAMPS AT VARIOUS EFFICIENCES.

candles in Fig. 49, the average duration of life is only 100 hours. In the first case, we should have had an 8-candle-power lamp lasting, say 18,000 hours, and representing a total output of 144,000 candle-hours, while in the second case we have a 32-candle-power lamp, lasting only 100 hours, and representing a total output of 3,200 candle-hours. In the first case, however, the candle-power would be obtained from the lamp at a comparatively heavy expense in energy, 2 1/2 times more energy, in fact, than that necessary for a candle in the second case. Moreover, the lamp in the first case would be very dull in color and unpleasant to the eye.

Between the above two extremes of long life and low efficiency, and short life and high efficiency, there exists a certain mean value most suitable for commercial purposes. This mean value, as lamps are now constructed, is in the neighborhood of 0.3 candle-per-watt, representing an average life time of 2,000 hours. This duration must, however, only be considered as the average lifetime, since it frequently happens that among a number of lamps manufactured by the same process, and with equal skill, some may only last about 100 hours at this efficiency, while others may greatly exceed 2,000 hours.

With any installation of electric lamps, it lies, therefore, in the power of the engineer in charge of the plant, to so operate the lamps that their life may be brilliant but short, or dull but long; as will depend entirely upon the pressure maintained at the lamp terminals. Thus, if a 16-candle-power incandescent lamp, intended for an activity of 50 watts, at 115

THE INCANDESCING LAMP. 187

volts, and therefore, to be operated at a normal activity of $\frac{16}{50}$ or 0.32 candle-per-watt, be steadily operated at a pressure one volt in excess, or 116 volts, its initial candle-power will be raised to 17 candles, but its lifetime will be abbreviated about seventeen per cent. Again, if the pressure be steadily maintained at 2 volts excess, or 117 volts, the initial candle-power will be normally 18 candles, but the probable lifetime will be reduced about thirty-three per cent.

When lamps are placed in somewhat inaccessible positions, such as near the ceilings of high halls, where there is some inconvenience in reaching them, it becomes particularly objectionable to have to renew these lamps too frequently. In order to avoid this, the plan is sometimes adopted of operating the lamps at a comparatively low efficiency and brilliancy, thus necessitating some extra expense in power, but greatly prolonging the average lifetime.

Fig. 51 represents the rate of variation of candle-power in similar samples of the same type of lamp when operated at different steady pressures. Curve No. 5 represents a 108-volt, 16-candle-power lamp of 0.286 candle-per-watt normal efficiency, operated at 108 volts. The candle-power slightly rises to about 16.6 candles, in 90 hours, and finally falls to 11 candles after 500 hours. The efficiency of the lamp at this time will, of course, be materially reduced.

Curve No. 4 of Fig. 51, represents the behavior of a similar lamp operated at 110 volts pressure. Here the initial candle-

THE INCANDESCING LAMP. 189

power is raised to 17 1/2 candles and after 500 hours burning, the candle-power is

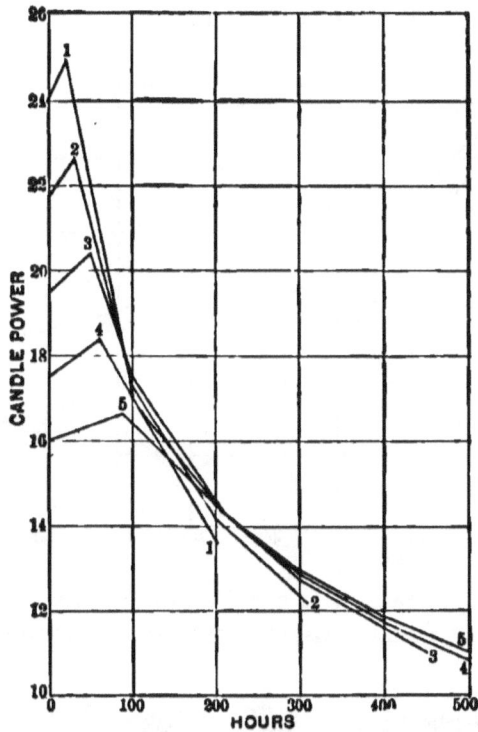

FIG. 51.—CURVES OF CANDLE-POWER IN LAMPS OF SAME TYPE OPERATED AT DIFFERENT FIXED EFFICIENCIES.

about 10 3/4 candles. Curve No. 3 shows similar results for a pressure of 112 volts.

Curve No. 2 for 113 volts and Curve No. 1 for 114 volts. Here the candle-power

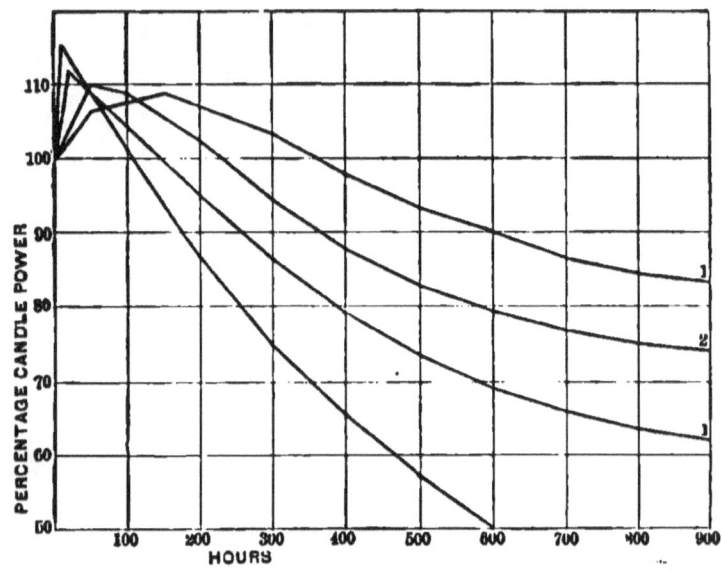

FIG. 52.—CURVES OF CANDLE-POWER OF THE SAME TYPES OF LAMP OPERATED AT DIFFERENT EFFICIENCIES.

commences at 24, but is less than 14 after 200 hours.

Fig. 52 represents the behavior of four similar types of lamps, operated steadily

THE INCANDESCING LAMP. 191

at various efficiencies, instead of at various pressures. Curve No. 1, represents the behavior with 0.25 candle-per-watt; Curve No. 2, 0.286 candle-per-watt; Curve No. 3, 0.333, and curve No. 4, 0.4 candle-per-watt. It will be seen that the high efficiency lamp falls to fifty per cent. of its initial candle-power in 600 hours, while the lowest efficiency lamp only loses ten per cent. of its candle-power in the same time.

It is evident, therefore, that no matter what the initial efficiency of a lamp may be, a time will come in its life when its efficiency must be low. When this point is reached, the amount of activity absorbed by the lamp, at constant voltage, is less than it was at the outset, seeing that the resistance of the attenuated filament is increased. On the other hand, the efficiency, or candles-per-watt, has diminished, and as regards its light, the electric power is more wastefully applied. It becomes, therefore, a question whether it would not be advisable to discard or break the lamp, and replace it by a new one, having a greater efficiency. By so doing we incur the expense of a new lamp earlier than if we waited for the old lamp to break naturally, but we utilize the power of the central station more economically.

The question of the *smashing point* of a lamp, or the point in the life of a lamp at which it may be deemed more economical to replace it by a new one, or its economical age, may be considered from three distinct standpoints; namely:

(1) From the central-station point of view.

(2) From the consumer's point of view.

(3) From the isolated-plant point of view.

The central station has usually to replace broken or useless lamps, and since the charge for service is based upon the ampere-hour, or the watt-hour, so long as the lamps burn and the consumer is fairly satisfied, the smashing point may be indefinitely extended. It is to the station manager's advantage, however, to maintain a steady pressure over the system of mains, so that the lamps are not forced above candle-power and their lives unduly abridged. In the best central stations great care is, therefore, always taken that the pressure is maintained as closely as possible to the normal. Should the pressure become markedly increased, the cost of renewing lamps will be rapidly augmented. Should it become markedly diminished, the consumers would be dissatisfied.

From the consumer's point of view the lamps would require to be operated at a high efficiency regardless of their lifetime, since he would thus obtain, from the power for which he pays, the highest brilliancy and the maximum quantity of light. Moreover, lamps which will not break before becoming seriously dulled, thus necessitating their renewal, are a disadvantage. From the standpoint of the consumer, therefore, the smashing point of a lamp is reached as early as possible, or at the opposite extreme to that of the station manager.

From the standpoint of the owner of the isolated plant, who is both producer and consumer, the smashing point will necessarily occupy some intermediate position;

for, while, on the one hand he desires to obtain the greatest amount of light for the activity produced or coal consumed, on the other hand he wishes to reduce the cost of lamp renewals as far as possible. No precise rule, however, can be laid down.

There are in general, two purposes for which light is ordinarily employed; namely, for actual use, as in reading, or other work, and for the æsthetic purposes of ornamentation. Regarding the latter purpose as a luxury, lamps may be changed as often as taste may dictate, but high-efficiency, high-brilliancy, short-lived lamps will be preferable. On the other hand, so long as a lamp fulfills the utilitarian purpose of enabling work to be conveniently and healthfully performed, it is waste of money to throw the lamp away. The smashing point, therefore of lamps intended to illumine working rooms depends largely upon the total candle-power installed. If this is ample in the first instance, a very considerable falling off in candle-power may be permitted without interfering with the usefulness of the light, and economy is rarely pushed to such a degree as to limit closely the candle-power installed for purposes of reading or working.

It is found, however, that in central station practice the best commercial results are secured with ample satisfaction to the consumers, if the efficiency at which the lamps are operated on the system, is such that the cost of lamp renewals is approximately fifteen per cent. of the total operating expenses of the station. Where the pressures between the mains can be closely regulated, it is preferable to employ high-efficiency lamps, but where the reverse is the case, low-efficiency lamps are desirable, since a low-efficiency lamp can stand an accidental increase in pressure with less detriment to its length of life than a high-efficiency lamp; for, being normally operated at a lower temperature, an increase of temperature may not be dangerous.

CHAPTER X.

LIGHT AND ILLUMINATION.

There are two technical words which are very apt to be confused in their meaning, and misused in their application; namely, "light" and "illumination." When correctly used, the word *light* signifies the flow or flux of light emitted by a luminous source, irrespective of the surface on which it falls, while the word *illumination* means the quantity of light received on a surface, per unit of area, whether received directly from the luminous source, or indirectly by diffusion and reflection from surrounding bodies. The words "light" and "illumination" are unfortunately often used synonymously, whereas it is evident that they denote distinct ideas.

By the *candle-power* of a source of light, we mean the luminous intensity of the source as measured in units of luminous intensity. The *unit of luminous intensity*, commonly employed, is the *British candle*, and is equal to the intensity of light produced by a candle of definite dimensions and composition, burning at the rate of 2 grains, or 0.1296 gramme, per minute. If we speak of a source of light as having, say a luminous intensity of 20 British standard candles, we mean that if that source were reduced to a mere point, it would yield as much light as 20 standard candles all concentrated at a single point.

A *standard of luminous intensity* very

generally adopted, except, perhaps, in English-speaking countries, is the *French standard candle*, called the *bougie-decimale*, or the 1/20th of the intensity of an international standard unit called the *Violle*. The Violle is a unit of luminous intensity produced in a perpendicular direction, by a square centimetre of platinum, at the temperature of its solidification. The British standard candle is slightly in excess of the bougie-decimale, one British candle according to Everett being 1.012 bougie-decimale.

If we imagine a point source of light, of unit intensity; *i. e.*, one standard French candle, or bougie-decimale, to be placed at the centre of a hollow sphere, of the radius of one metre, or 39.37", then the total internal surface of the sphere will receive a definite total quantity of light. Each

LIGHT AND ILLUMINATION.

unit of area, a square metre of interior spherical surface, will receive one unit of light; and, since the total interior surface of such a sphere contains 4 × 3.1416 = 12.566 square metres, the total quantity of light received on the interior surface will be 12.566 units, called *lumens*. Consequently, if a point source of unit intensity emits 12.566 lumens, a source, of say 20 bougie-decimales, would yield an intensity of 251.32 lumens.

It is obviously not necessary that the surface, which surrounds the unit point source, should be spherical. The same amount of light; namely, 12.566 lumens, will be given off by the source independently of the shape of the receiving surface, so that when such a source is placed, for example, alone in a room, the total quantity of light which falls directly upon the walls, ceiling and floor of the room from the source, will be 12.566 lumens. This will be true in fact if there are other sources of light in the room at the same time. The quantity of light which each point source will emit will be 12.566 times its luminous intensity.

If one lumen falls perpendicularly and uniformly over the surface of one metre, the *illumination* of that surface will be one lux, the *lux* being the *unit of illumination*. If, for example, a bougie-decimale is located at the centre of a sphere of one metre radius, then each square metre of the interior surface of the sphere will receive one lumen of light, and this light falls everywhere perpendicularly upon its surface. The illumination on the interior surface is everywhere one lux. A bougie-decimale produces, therefore, when acting alone, an illumination of one lux at a distance of one metre.

The mistake is not infrequently made, that because a surface receives light directly from a given source of known intensity, its illumination can be determined by mere calculation of its distance from the source. It must be remembered that it also receives reflected or diffused light from all neighboring surfaces, which, consequently, tend to increase its illumination.

The *law of illumination* from a single point source, acting alone; *i. e.*, in a space where all reflected or other light is excluded, is that the illumination varies inversely as the square of the distance from the source. Thus, we know that a bougie-decimale produces an illumination of one lux upon a surface held perpendicularly to the rays of light at a distance of one metre. If the surface be held at a distance of 2 metres from the point source, in a room otherwise dark, the illumination will be 1/4th of a lux, or $(1/2)^2$; at a distance of 3 metres the illumination would be similarly 1/9th lux, or $(1/3)^2$. Generally, a point source, of say 50 bougie-decimales, would produce, at a distance of say 5 metres, an illumination of $\frac{50}{5^2} = 2$ luxes.

In practice, if a surface were placed in a room at a distance of 5 metres from a source of 50 bougie-decimales, the illumination received, if the rays be allowed to fall perpendicularly upon the surface, will be more than 2 luxes, because this amount of illumination will be produced by the direct action of the source in a space from which all other light was excluded, whereas, reflection from the walls of the rooms, mirrors, ceilings, etc., will increase this amount of illumination to, perhaps, 10 luxes, and, if it were possible to have the surfaces of the walls perfectly reflecting, the illumination which would be produced in all parts of the room would be indefinitely great. Consequently, the amount of illumination received upon the surface of a desk or table, depends not only upon the number of lamps in a room, on their candle-power and arrangement, but also upon the character of the surfaces of the walls and furniture. Therefore, the question of lighting a room is not altogether a question of its dimensions and of the total candle-power placed in it, but also depends upon the arrangement of the lamps, and the character of the decoration and furniture, and the nature of their reflecting surfaces.

The illumination required for comfortable reading is from 15 to 25 luxes on the surface of a printed page. Any illumination less than 10 luxes is fatiguing, if long continued. The illumination produced by full moonlight is about 1/8th lux, and that by full sunlight 80,000 luxes. The illumination in a street as ordinarily lighted by arc lamps, is, perhaps, 50 luxes near the ground below an arc lamp, and about 1 lux near the ground midway between the lamps.

The luminous intensity of an incandescent lamp is not the same in all directions, owing to the fact that the filament is not a point source, and that, in some positions, a greater surface area of filament is exposed than in others. Thus, a simple horse-shoe filament, with both legs in one plane, gives less light in this plane than in any other plane passing through the vertical, because each leg intercepts the light from the other. The *mean spherical candle-power* of a lamp, in bougie-decimales, is the quantity of light it emits in lumens, divided by 12.566. In other words, the mean spherical candle-power of a lamp is the equivalent point source which emits as much light in all directions as the actual lamp does. The spherical candle-power of an incandescent lamp is usually about twenty per cent. less than its *maximum horizonal intensity*, so that when we speak of a 16-candle-power lamp, a point source of about 13 candles intensity would supply the same total number of lumens as is emitted from the actual lamp. A point source would have no base or socket and would disperse light equally in all directions.

It may be of interest to note that a *lux-second*, that is to say, the total *time-illumination* produced by one lux for one second of time, has been accepted by the International Photographic Congress of Brussels as the *unit of time-illumination*, under the name of the *phot*. A phot is a *lux-second*, and is a unit of time illumination employed in photography. It is well known in photography that the *actinic* effect of light, that is its power of effecting chemical decomposition as utilized in photography, depends, for a given quality of light, both on its intensity and on the duration of its action. In photography, therefore, this practical unit was required to represent the product of illumination and time.

The candle-power of incandescent lamps varies from 1/2 candle up to 100 British Standard candles, although both larger and smaller candle-powers have been specially prepared. The 16-candle-power lamp is generally employed in the United States, and the 10-candle-power lamp is generally employed in Europe. The sizes usually manufactured are 1/2, 1, 2, 3, 4, 5, 6, 8, 10, 16, 20, 32, 50, 60 and 100.

CHAPTER XI.

SYSTEMS OF LAMP DISTRIBUTION.

BROADLY speaking, there are two general methods by means of which lamps may be connected with a generating source of electricity; namely, *in series,* so that the current passes successively through each lamp before it returns to the source, and *in multiple* or *parallel;* so that the current divides and a portion passes through each lamp.

Fig. 53, represents three lamps connected in series, and Fig. 54, represents three lamps connected in parallel. In the case of the series connection shown in Fig.

SYSTEMS OF LAMP DISTRIBUTION. 211

53, the three lamps are so connected that the current from the source, entering the line at +, flows in the direction indicated by the arrows, passes successively through the lamps A, B, and C, returning to the

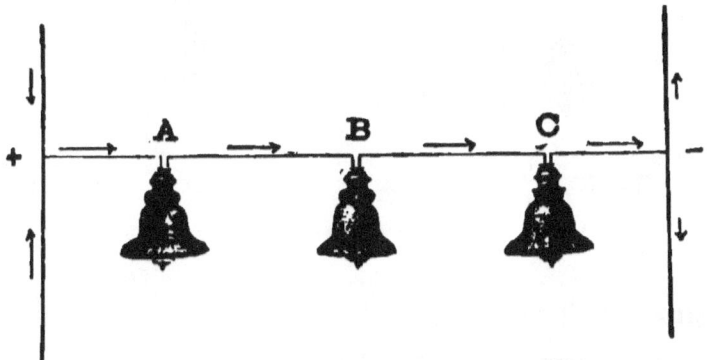

FIG. 53.—SERIES-CONNECTED LAMPS.

source at the negative end of the line. In Fig. 54, the three lamps A, B, C, are connected as shown, to the positive and negative leads respectively, and the current passes through them in the direction indicated by the arrows. The arterial system of the human body furnishes an example of a parallel or multiple system, since the blood flow divides into a very great number of different channels or capillaries, and, after passing through these independ-

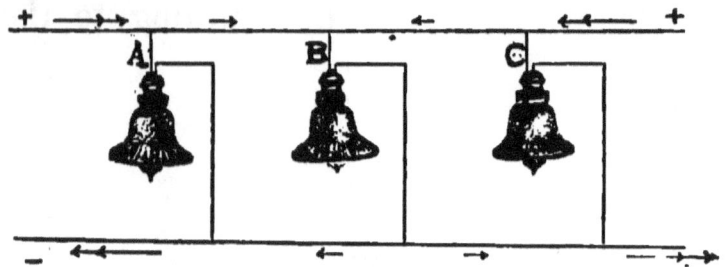

FIG. 54.—MULTIPLE-CONNECTED LAMPS.

ent channels, finally unites in the veins and returns to the heart.

In order to compare the relative advantages of the series and multiple methods of electric distribution, let us suppose that a house has to be lighted electrically at a distance of a mile from the dynamo, and that for this purpose an amount of light represented by 1,000 candles is required, distributed in 50, 20-candle-power lamps. Further, let us assume that the same efficiency is secured in the operation of these lamps, whatever system we may adopt, or whatever dimensions the lamp filament may take, a supposition in accordance with general practice. Let us suppose that this efficiency will be 1/4 candle per watt. We shall then require to expend 4,000 watts of electric energy in the lamp filaments in the house. This amount of activity might be electrically expended in a great variety of ways, as regards the pressure and current of delivery, but it will suffice to compare two ways only; namely, the delivery of 4 amperes at a pressure of 1,000 volts, and the delivery of 1,000 amperes, at a pressure of 4 volts. In each of these two cases the activity delivered will be the same; namely 4 KW.

If it be required to expend only 1,000 watts in the main conductors carrying the current to the house, when all the lamps are turned on, 5,000 watts or 5 KW will have to be supplied at the generator terminals, of which twenty per cent., or 1,000 watts, is permitted to be lost in transmission in the leads, as heat. We know that this loss will be the product of the current strength and the drop in volts in the two conductors. In the first plan of 1,000 volts and 4 amperes at the house, the drop of pressure in the two wires must be 250 volts, in order that 250 volts × 4 amperes = 1,000 watts, so that the resistance of the two wires together, which shall produce a drop of 250 volts, with a current of 4 amperes, will be 62 1/2 ohms, or 31 1/4 ohms to each wire one mile in length. The nearest size of wire to that which has a resistance of 31 1/4 ohms to

SYSTEMS OF LAMP DISTRIBUTION. 215

the mile, is, No. 18 B. & S. or A. W. G., having a diameter of 0.0403", or, approximately, the $\frac{1}{25}$th of an inch. Such a wire would weigh about 26 lbs. and the two wires forming the complete circuit would weigh about 52 lbs.

Considering the second plan of 4 volts and 1,000 amperes, the drop in the wires would have to be only one volt. In order that the activity expended in them should be 1 KW since 1 volt × 1,000 amperes = 1 KW the resistance in the two wires together, to permit of a drop of but 1 volt with 1,000 amperes, must be $\frac{1}{1,000}$th of an ohm, since 1,000 amperes × $\frac{1}{1,000}$th ohm = 1 volt. The two wires together must, therefore, have a resistance of $\frac{1}{1,000}$th ohm, or each must have a resistance of $\frac{1}{2,000}$th ohm. A wire which would have this resistance would have 62,500 times the cross-section and weight of the wire in the preceding case; so that, instead of requiring 52 lbs. of copper, in all, for the two miles of conductor, we should require approximately 3,250,000 lbs., or 1,625 short tons of copper.

It is, therefore, evident that although a given electric activity can be expended in incandescent filaments either at a high pressure or at a low pressure, the advantage of a high pressure is very great, when the power is to be transmitted electrically to a distance. If we double the pressure of transmission; *i. e.*, if we double the number of volts between the two main conductors, we require four times less cop-

SYSTEMS OF LAMP DISTRIBUTION. 217

per for a given percentage of loss of activity in them. Thus, in the preceding case, when we increased the pressure of delivery from 4 volts to 1,000 volts, we increased it 250 times, and, therefore, we diminished the amount of copper which was required for twenty per cent. loss, 250 × 250 or 62,500 times. Consequently, the first essential for economical distribution of electric power to a distance, either for lamps, or for any other purposes, is high electric pressure of delivery.

If then we adopt provisionally, the plan of supplying the house above considered at a pressure of 1,000 volts and 4 amperes, this would appear to be most readily carried out at first sight, by connecting 50 lamps in a single series through the house, each lamp being intended for 20 volts and 4 amperes. When all the lamps are lighted, the total current would be 4 amperes, when the total pressure of delivery amounted to 50 × 20, or 1,000 volts. Though such a system could, doubtless, be installed, yet it would possess several disadvantages. In the first place, unless some device were provided whereby a faulty lamp became automatically short-circuited, the failure of any single lamp would interrupt the entire circuit and extinguish all the other lamps. Moreover, the pressure which would have to be supplied to the house between the main conductors would depend upon the number of lamps employed. When a single lamp only was lighted, the pressure required would be 20 volts and the current 4 amperes. This would mean that the generating dynamo in the station supplying the house could only be used for that particular house, since, if two houses were supplied from

SYSTEMS OF LAMP DISTRIBUTION.

the same dynamo, one might have all the lamps turned on and thus require 1,000 volts and 4 amperes, while the other might have only half its lamps turned on, thus requiring 500 volts and 4 amperes. For this and other reasons it is now universally considered that incandescent lighting on any extended scale must necessarily be conducted by a multiple-arc system.

It would be practically impossible to construct incandescent lamps capable of being operated in parallel at the pressure of 1,000 volts assumed in this case; for, each lamp would have to be capable of taking a current of $\frac{4}{50}$ths ampere, at a pressure of 1,000 volts. The resistance would have to be 12,500 ohms hot, so that the filament would have to be very fine and long. Such a lamp would be quite impracticable, and, moreover, the pressure of 1,000 volts is not considered safe to introduce into a building. The maximum pressure for which it has been possible, until recently, to construct incandescent lamps has been 120 volts, so that incandescent lighting between a single pair of conductors has been practically limited to a pressure of 115 volts at the lamp terminals.

The lack of economy of 115-volt pressures for incandescent lighting at a distance, as regards the amount of copper required, was early apparent, and a method was invented and introduced which is practically a compromise between the series and parallel systems; that is to say a method was invented whereby the advantages of the parallel connection of lamps were secured, together with the

SYSTEMS OF LAMP DISTRIBUTION. 221

advantage of higher pressure obtained by coupling lamps in series. This method is fundamentally what is called a *series-multiple system*, and in practice is what is generally called the *three-wire system*.

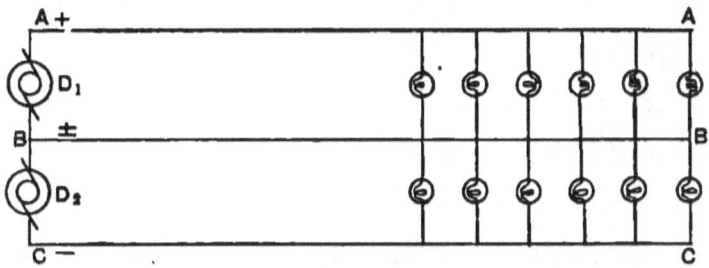

Fig. 55.—Three-Wire System, Series-Multiple Connection.

The three-wire system is illustrated in Fig. 55. Here there are two multiple-arc circuits, one between the mains A and B, and the other between the mains B and C. Between each of these pairs of mains the pressure is 115 volts, supplied by a separate dynamo as shown. The positive terminal of the dynamo D_2, being connected to the negative terminal of the dynamo D_1, it is evident that between the mains A A and C C, there will be a pressure of 230 volts. In the case shown there are 12 lamps in all, or 6 on each side of the system, so that about 3 amperes will be flowing through the mains A A and C C, and no current will pass through the neutral conductor B B. If the number of lamps on the two sides of the system be unequal, the difference between the current strengths will return by the neutral conductor; for example, if all the lamps between B and C, are turned off, 3 amperes will flow along the positive main A A, and return by the neutral main B B, no current passing through the negative main. Since, on the average, when the wiring is judiciously carried out, the loads on the two sides of the system will usually nearly balance each the outside mains, and may therefore be much lighter. If the neutral conductor could be entirely dispensed with, we know that the copper required to supply the system with a given percentage of loss in transmission would be four times less on the three-wire system, than on the two-wire system, since the pressure of distribution would be doubled. The three-wire system would, therefore, ensure a saving of seventy-five per cent. in the amount of copper required for the mains. In practice, however, the amount of copper in the neutral conductor averages, over an entire system, about sixty per cent. of that in the outside conductor, so the neutral is made a little more than half as heavy as either of the outside mains. Under these conditions, the actual saving in weight of copper throughout the supply conductors of a three-wire system is about 67 1/2 per cent. over that necessary for a two-wire system having the same loss in transmission.

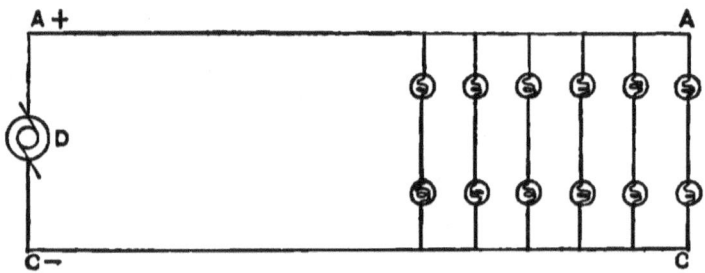

FIG. 56.—MULTIPLE-SERIES CONNECTION.

Fig. 56 represents a multiple-series system equivalent to a three-wire system with no neutral, and supplied at 230 volts pressure. The disadvantage of such a system, however, is that two lamps have to be turned on and off simultaneously. The three-wire system of Fig. 55 gives independent control over every lamp.

SYSTEMS OF LAMP DISTRIBUTION.

The principle of the three-wire system has been extended to four- and five-wire systems. Four-wire systems are very rare. Five-wire systems are employed in Europe but have never come into favor in the United States. A five-wire system saves about ninety per cent. of the copper required for a two-wire system, but requires four dynamos in series at the central station, five sets of conductors and complication in house wiring and meters.

The three-wire system is in very extended use in the United States. It commonly happens that one three-wire central station will distribute light and power over an area whose radius is somewhat greater than one mile, whereas, without the use of the three-wire system, the radius of commercial incandescent lighting from a central station would be probably only about one-half a mile or eight times less area.

The drop of pressure, which is permitted in incandescent lighting, does not depend entirely upon the activity uselessly expended in the main conductors. For example, in cases where capital would be difficult to secure, and the interest upon the capital invested would be large, it would be desirable to employ comparatively small conductors, and waste a comparatively large percentage of the total power in them. This would necessitate a comparatively great difference of electric pressure between the lamp terminals and the generator terminals. In the case of a single house supplied with 115-volt lamps, it would not be a matter of much consequence whether the pressure at the central station were 116 or 166 volts, provided the lamp pressure remained constant, but where incandescent lamps are distributed along street mains, in a city, and have to be supplied at all distances from a few yards to a mile or more from the central station, it is absolutely necessary that the pressure shall be nearly uniform throughout the system, since, otherwise, the lamps in or near the station will be at an unduly high pressure, and will consequently be brilliant and short-lived, while the more distant lamps will be burned at an unduly low pressure, and be dull and long-lived. The drop of pressure permissible in the supply conductors is, consequently, as much a matter of regulation of pressure, and of uniformity of candle-power, as it is a consideration of economy in the transmission of electric power. If lamps were less sensitive to changes in pressure than they are, the amount of copper which would be employed in incandescent lighting would be less than it actually is, but all the improvements made of recent years in incandescent lamps have been improvements in their efficiency, whereby a smaller amount of activity is required for a given production of light, and this, as we have seen, is attended by the development of a higher temperature and a greater sensibility to variations in pressure, so that the most economical lamps are also lamps which, other things being equal, require a closer regulation of pressure at their terminals.

The difficulty of maintaining a nearly uniform pressure over all parts of the mains of a large incandescent system has been largely overcome by the use of what are called *feeders*. A feeder differs from an ordinary supply conductor, or main, in that no lamp

SYSTEMS OF LAMP DISTRIBUTION. 229

or receptive device is directly connected with it; its sole purpose being to supply the mains from the central station at some distant point as indicated in Fig. 57. Here D, is the dynamo at the central station. $F\ F$, are feeders carrying the current from the dynamo to some cen-

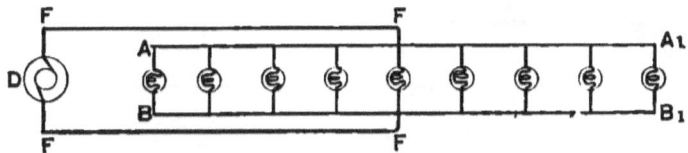

FIG. 57.—FEEDER DISTRIBUTION.

tral point in the mains $A\ A_1$, $B\ B_1$. In this way the difference in pressure between the various lamps depends only on the drop of pressure in the mains, and not on the drop of pressure in the feeder. Thus, if the pressure at the lamps at A and A_1, be 115 volts, the pressure at the feeding point F, may be 116 volts, while the pressure at the dynamo may be 150 volts.

230   ELECTRIC INCANDESCENT LIGHTING.

If the same lamps were supplied without feeders as shown in Fig. 58, and the same limiting difference of pressure maintained between the lamps as in Fig. 57, namely 1 volt, it would be necessary to have practically 116 volts at the dynamo terminals and 115 volts at the most distant lamp.

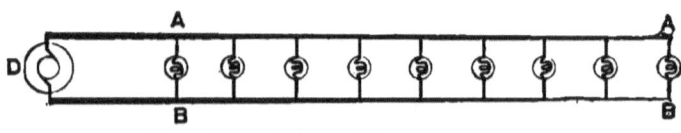

Fig. 58.—Tree Distribution.

This probably would require much more copper in supply conductors than when feeders are employed. The conductors in Fig. 58 are shown as tapering or diminishing in size towards the distant end.

Feeders may equally well be applied to three-wire systems. Fig. 59 represents diagrammatically the supply-mains of a city district containing four blocks, 1, 2, 3

SYSTEMS OF LAMP DISTRIBUTION. 231

and 4. Here the three-wire mains extend round the block-facings in one continuous network. These mains are supplied from the central station at *S*, by the three-wire

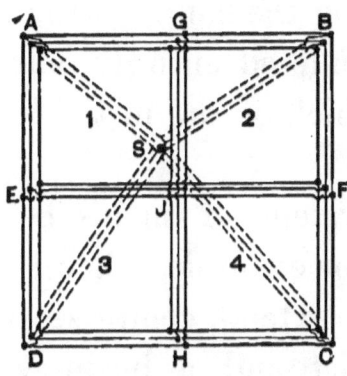

FIG. 59.—DIAGRAM OF THREE-WIRE FEEDER DISTRIBUTION.

feeders represented by the dotted lines, at the feeding points, *A*, *B*, *C* and *D*. In this way the pressure in the network of mains may be within two per cent. of the mean value, of say, 115 volts, while the pressure at the central station may be 130 volts, representing a drop in the feeders of

15 volts. If the lamps were connected across the feeders they would be subjected to a total difference of pressure, over the entire system, amounting to 17 volts, but, by connecting the lamps to the mains only, they are rendered entirely independent of the drop which occurs in the feeders.

If the system of mains be unequally loaded, as for example, when the area over which they extend, comprises both a residence district and a business district, so that the load shifts in the morning and afternoon to the business district, and in the evening, almost entirely to the residence district, it may happen that the load on some particular feeders may be much greater than the load on others. Consequently, the drop in the loaded feeders will be in excess of that on the comparatively idle feeders. Under these circumstances, the pressure at the mains, near the terminals of the idle feeders, will be higher than that at the terminals of the loaded feeders, thus bringing about an inequality of pressure, prejudicial to the life and proper performance of the lamps.

The difficulty arising from the inequality in the feeder load may be overcome in one or more of four ways:

(1) By disconnecting certain feeders from the *bus-bars*, or main terminals in a central station, so as to increase the load and drop on the remaining feeders.

(2) By introducing artificial resistances, called *feeder regulators*, into the circuit of the idle feeders, so as to increase artificially the drop of pressure which exists in them.

(3) By employing more than one pressure in the central station, that is to say, by having one set of dynamos operating at

Fig. 60.—Feeder Equalizer Resistance.

supply of the shorter feeders to the area within the vicinity, and another set of dynamos delivering, perhaps, 135 volts, for

FIG. 61.—EQUALIZER SWITCH.

the supply of longer feeders connected to the outlying districts.

(4) By introducing more copper into the system, either in the form of additional

feeders, so as to share and equalize the load, or in the form of more numerous mains to distribute and equalize the pressure.

Fig. 60, represents a form of resistance, suitable for feeder regulation. Here a number of spirals of heavy iron wire are mounted in a fire-proof frame and so arranged that under the influence of the handle and switch, shown in Fig. 61, they may be inserted in the circuit of a feeder either in parallel or in series.

In modern large central stations feeder equalizers are rarely employed. The best practice employs more than one pressure.

CHAPTER XII.

HOUSE FIXTURES AND WIRING.

The incandescent lamp, when located in a house, is either installed as a fixture, or a certain freedom of motion is given to it, so that, within certain limits, the lamp is portable. This portability is effected by maintaining the lamp in connection with the mains by means of a flexible conductor or lamp cord, usually called a flexible cord. The lamp is then portable to the extent of the length of the cord. Various forms are given to portable lamps, two of which are shown in Figs. 62 and 63. In Fig. 62, the flexible cord $c\ c$, is attached to a reading lamp, mounted on a stand as shown.

238   ELECTRIC INCANDESCENT LIGHTING.

FIG. 62.—PORTABLE LAMP FOR DESK USE.

This lamp can be raised and lowered within a limited range, as well as turned

Fig. 63.—Flexible Lamp Pendant with Adjuster.

about its axis without shifting the base. Fig. 63, shows a form of movable lamp, in which a limited portability is obtained by what is generally known as a *flexible pendant.* Here the lamp is hung from

Fig. 64.—Flexible Support for Lamp.

the ceiling by a flexible lamp cord. By means of an adjuster J, the lamp can be raised or lowered.

The limited portability given to a lamp, by attaching it to a sufficiently long flexible pendant, enables the light to be applied to a variety of purposes, such, for example, as the lighting of a music stand,

Fig. 65.—Flexible Support for Desk Lamp.

as shown in Fig. 64, or the lighting of a desk, as shown in Fig. 65. Fig. 66, shows a device for tilting a flexible pendent lamp in any desired direction.

Fixed lamps, as their name indicates, are lamps attached to electric fixtures, and,

242   ELECTRIC INCANDESCENT LIGHTING.

therefore, cannot be moved. They take a great variety of forms, such as the bracket lamp shown in Fig. 67, designed for

FIG. 66.—TILTED LAMP.

attachment to the wall. For ceiling attachment, lamps are either made of the simple pendant type, as shown in Fig. 68, or several lamps are placed together in a cluster in an *electrolier*, as shown in Fig. 69. As in gas lighting, the incandescent bracket lamp is sometimes given a movable arm so as to permit the lamp to

FIG. 67.—BRACKET LAMP.

be moved in one plane, within a certain radius. Such a lamp is shown in Fig 70.

The size of the wire employed inside a house will depend upon the amount of current which the conductor is designed to

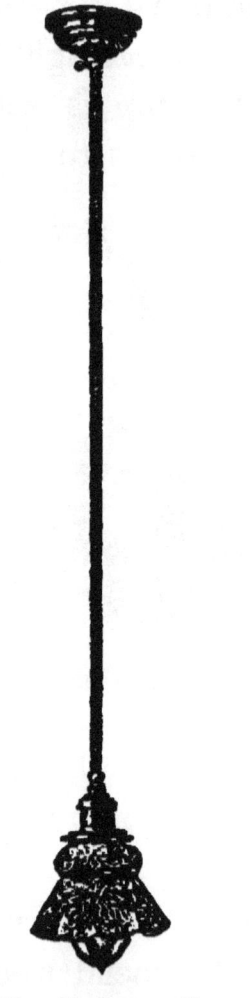

Fig. 68.—Pendant Lamp.

carry. When of small size, the conductor is given the form of a single wire, but,

FIG. 69.—ELECTROLIER.

in order to secure greater flexibility, larger sizes are almost invariably *stranded*, that is, composed of several independent wires. The former are called *solid wires* and the latter *stranded wires*. In Fig. 71, the solid conductor is marked c, and has one coating of insulator d, which is afterward

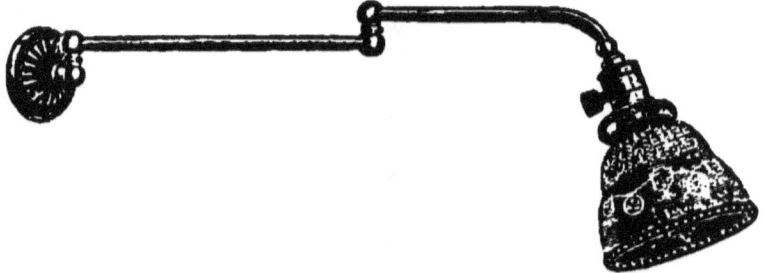

Fig. 70.—Bracket Lamp with Movable Arm.

covered by a braiding b. The stranded conductor shown at C, consists of seven wires, twisted together as shown, and is covered by two coatings of insulating material, D and E, respectively, and finally by a coating of braid B.

Fig. 72, shows two other forms of stranded conductors; the wire marked A,

HOUSE FIXTURES AND WIRING. 247

is provided with a highly insulating material called *okonite;* that marked *B*, has in addition, a coating of braid. The wire at *A*, is equivalent to No. 6 A.W.G.

Fig. 71.—Solid and Stranded Conductors.

in cross-sectional area. The insulation resistance of a mile of this wire, when submerged in water, is 1,000 million ohms, that is, one billion ohms, or a begohm.

A flexible cord, such as has already been referred to in connection with portable lamps, is necessarily a double conductor, since the current must be passed both into and out of the lamp. These two conductors are separately insulated, and are then either twisted together, forming what is called a *twisted double con-*

FIG. 72.—OKONITE-COVERED STRANDED WIRES.

ductor, or are laid side by side, and laced together by a covering of braid, forming what are then called *parallel* or *twin conductors*. Fig. 73, shows some forms of double flexible conductors. These are first separately insulated and are then *silk-covered*. They are sometimes technically called *silk lamp cord*. In order to attain the flexibility required in such cords they are composed of a comparatively large number of fine copper wires stranded together.

Fig. 73.—Forms of Double Flexible Conductors.

We will now trace the network of conductors in a house which we will suppose receives its current from the street mains, to the lamps in the different portions of the house. First; proceeding from the street to the house, we find a set of conductors leading into the house, called the *service wires*. These in a two-wire system consist of two conductors, and in a three-wire system of three conductors. Within the house the system of conductors may be arranged under the following heads; namely,

(1) The *risers*, or the supply wires which carry the current up from the service wires to the different floors of the house. They may be a single set or a multiple set, but each set will be double or triple according as the house is wired on the two- or three-wire system.

(2) The *mains*, or the principal supply conductors running from the risers or service wires along the different corridors or passages. There are usually as many separate systems of mains as there are floors.

(3) The *sub-mains* or the *supply-conductors* which branch off from the mains along the side passages.

HOUSE FIXTURES AND WIRING

(4) *Branches or taps carried* from the mains into the rooms or to fixtures in halls. Roughly speaking, therefore, the risers correspond to the trunk of a tree through which the sap is fed; the mains correspond to the boughs; the sub-mains to the smaller boughs; and the branches to the twigs. The lamps or fixtures correspond to the leaves and flowers.

Risers are usually of larger cross-section than the mains; the mains are of larger cross-section than the sub-mains, and the sub-mains, in their turn, are larger than the branches. This must naturally be the case, since the risers must carry all the current, the mains divide the current among themselves, and the branches carry only the current of the few lamps wired upon them. We may imagine that each lamp has a certain size of wire connected with it from the street mains, direct to the socket, and, that since these wires run side by side, they may be regarded as collected into a single larger wire, such as a main or riser.

The smallest size of wire which is permitted to be used in wiring a building, in the United States, is No. 18 A. W. G. wire, having a diameter of 0.040".

The wiring of a building should be designed in such a manner, that when all the lamps are burning at any one time, the drop in pressure between the street mains and the most distant lamp shall not exceed a certain small percentage, usually three per cent. This drop is calculated by determining the total amount of current in amperes which will pass through the various mains, sub-mains and branches, determining for each such a resistance as will, when carrying this current, produce drops of pressure, the maximum sum of which along any line shall not exceed the required percentage.

In very large buildings, a *feeder system* is sometimes employed; that is to say, the service wires are connected by feeders to *centres of distribution*, from which mains extend both upward and downward.

Two supply wires, carrying the full lamp pressure between them, are never permitted to remain in contact with each other, even though both are insulated, except in cases of flexible conductors, which are in plain view and which never carry, under normal circumstances, a powerful current.

Insulated wires are never allowed to come into contact with concealed woodwork, but when passing through wooden beams or floors should be protected by insulating tubes of porcelain or hard rubber.

There are three methods of carrying out interior wiring; namely,
 (1) Cleat work.
 (2) Moulded work.
 (3) Concealed work.

Cleat work is the simplest and cheapest, but least ornamental type of wiring. The wires are carried in insulating receptacles, of wood or porcelain, in plain view on the ceilings or upper part of the walls. The wires should never rest directly upon the walls or ceilings, but should be supported by the cleat, a full half inch away from the same.

Fig. 74, shows two forms of wooden *cleats*. *F, F,* are the front pieces which

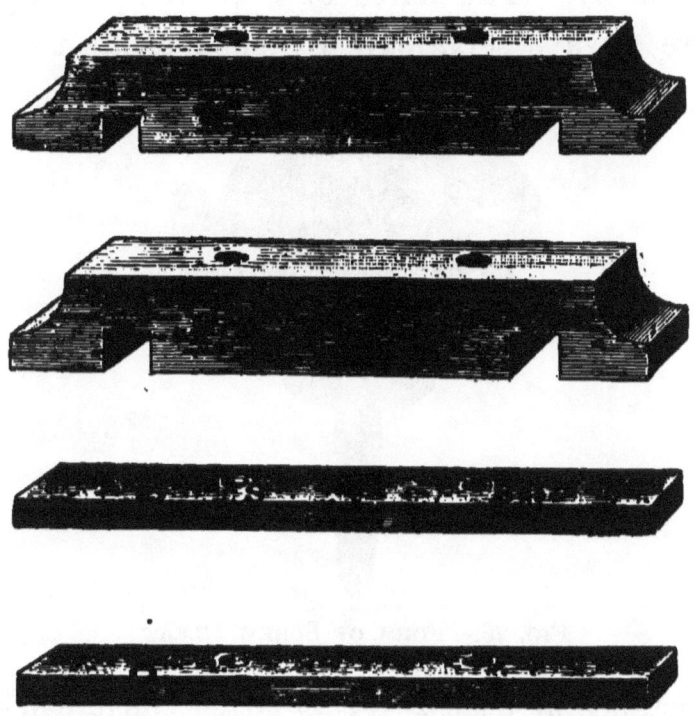

FIG. 74.—WOODEN CLEATS.

clamp the wires, and *B B,* are the brackets which support the wires free from the walls. *S S*, are the screw holes by which the cleats are secured in place and clamped together. The wires of opposite polarity are always separated by such cleats to a

FIG. 75.—FORM OF SCREW CLEAT.

distance at least 2 1/2" apart. Cleat work is only suitable for indoor work in dry localities. Figs. 75 and 76 show forms of screw cleats employing respectively wood and glass insulation around the wire.

HOUSE FIXTURES AND WIRING. 257

Moulded work is more expensive than cleat work, but presents a more sightly appearance. Fig. 77, shows different forms of *three-wire moulding*, suitable for different

FIG. 76.—FORM OF SCREW CLEAT.

sizes of conductor. The moulding is made in two parts; namely, the base or *mould*, with the grooves formed in it, and the upper part, or *capping*, which covers the mould. These moulds and cappings are

258   ELECTRIC INCANDESCENT LIGHTING.

usually made of soft pine, and are cut into lengths of about ten feet. The moulds are

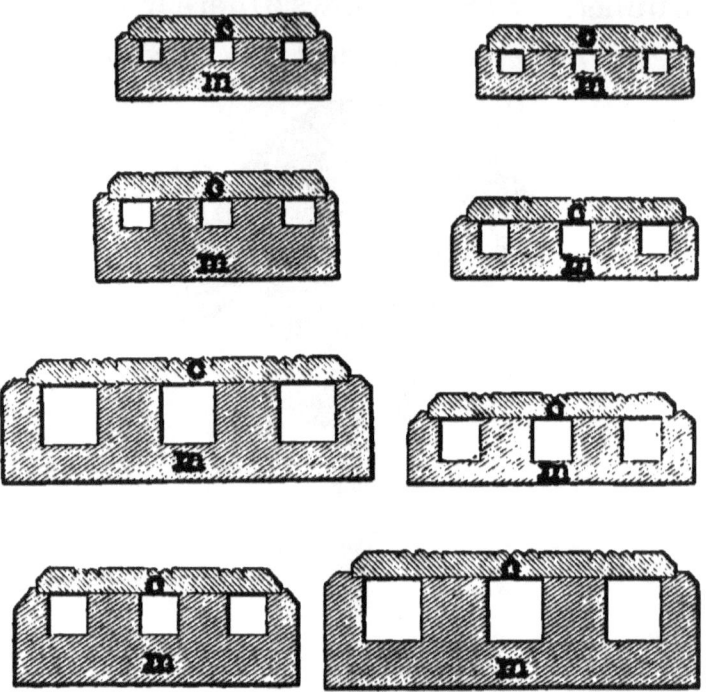

Fig. 77.—Sections of Mouldings.

first screwed in position on the walls or ceilings; the wires are then laid in them, and, finally, the cappings are secured by screws over them, care being taken not to injure the conductors in screwing on the cappings. The mouldings are usually

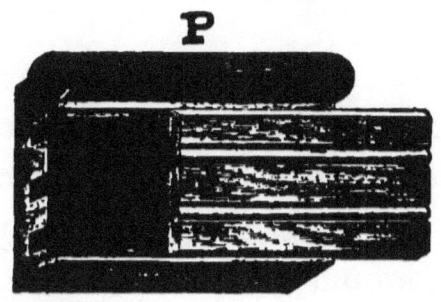

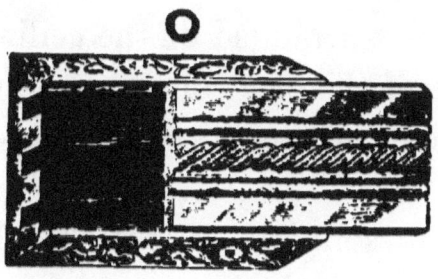

Fig. 78.—Picture and Ornamental Moulding.

painted of a color to conform with the ornamentation of the walls or ceilings on which they rest. Fig. 78, represents at P, a form of moulding suitable for hanging pictures around the walls of a room, and at *O*, a type of ornamental moulding.

The most difficult problem connected with the distribution of wires by moulding, lies in the connection of the electroliers with the wires in the passages without presenting an unsightly appearance. This is sometimes accomplished by the use of *dummy moulding*, or ornamental mouldings symmetrically arranged on the ceiling from the centres of the electrolier, in one only of which mouldings the wires are placed.

The best solution of the problem of avoiding the unsightly appearance of wires is obtained by *concealed work*, where the conductors are buried under floors or in the walls and ceilings. This method should not be adopted unless the wires besides

HOUSE FIXTURES AND WIRING. 261

their insulating cover, are provided with a moisture-proof sheet or tube of *papier-*

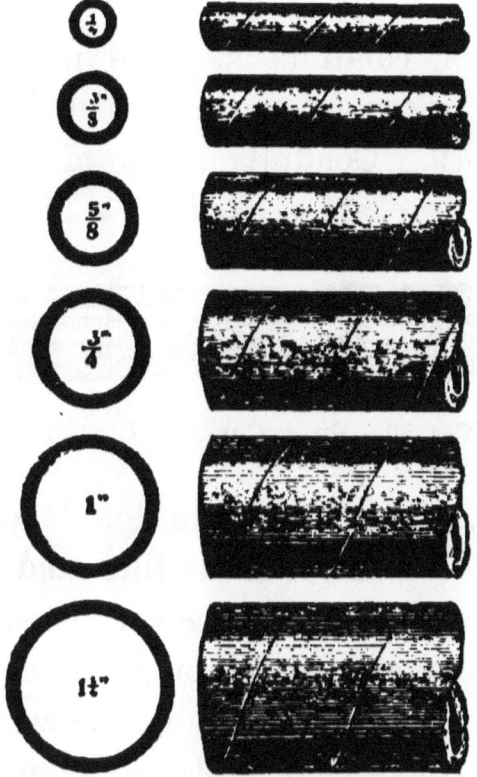

Fig. 79.—Interior Conduits.

maché or metal. Such protecting tubes form in reality a conduit employed inside the building and generally called an *interior conduit*.

Interior conduits may be made in a variety of ways, one of which is shown in Fig. 79. Conduits made of tubes of *papier-maché* are soaked in a bituminous

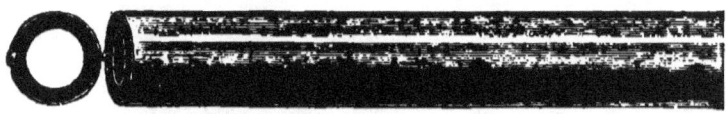

Fig. 80.—Brass-Covered Conduit.

solution, which serves the double purpose of rendering them insulating and practically water-proof. Moreover, when house wires are carried through a complete system of interior conduits, the wires can be withdrawn and replaced at any time without disturbing the walls or ceilings. Fig. 80, shows a conduit tube sheathed with a thin layer of brass so as to be water-tight. Fig. 81, shows a brass tube joint connecting different lengths of conduit. *A*, is a joint for connecting ordinary conduit, and *B*, a joint connecting brass-covered sheathed conduit. Where several

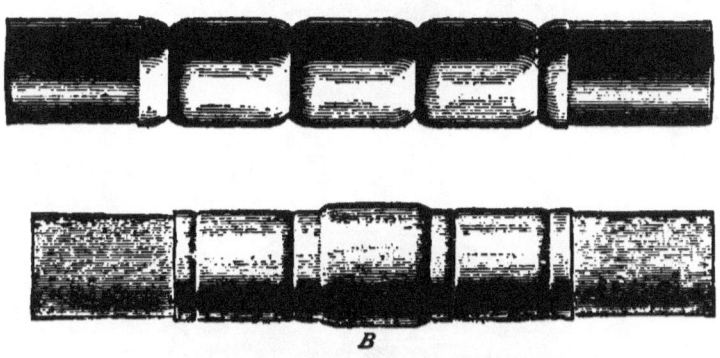

Fig. 81.—Interior Conduit Joints.

conduits are united together, a *junction box* is provided containing a number of outlets, corresponding with the number of conduits as shown in Fig. 82. These boxes are closed by a metallic cover. Fig. 83, shows two forms of *connecting boxes* for holding

264   ELECTRIC INCANDESCENT LIGHTING.

joints between the mains and branches of a two-wire or three-wire system running in

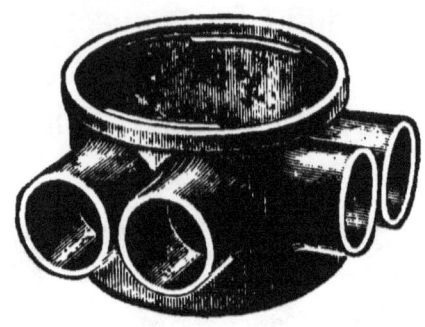

FIG. 82.—INTERIOR CONDUIT JUNCTION BOXES.

interior conduits. The branch wires in this case proceed from the box through the smaller apertures. In cellars the conduits are frequently heavily sheathed with brass or iron.

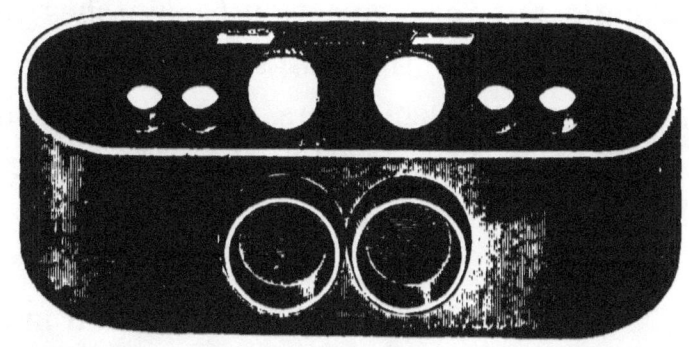

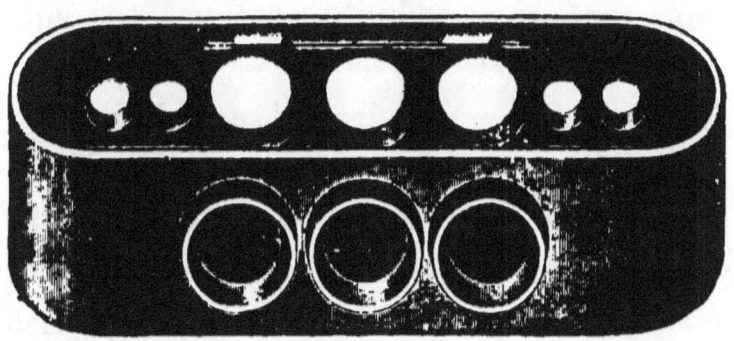

FIG. 83.—CONNECTION BOX FOR INTERIOR CONDUITS.

The cost of wiring a building depends in a great measure upon the number of *outlets* required, that is on the number of points at which the wires must be brought out for connection to switches and lamps. Where the lamps are installed in groups, as in electroliers, fewer outlets are required and the cost is less than if the lamps were installed singly.

Concealed work is generally much more readily and cheaply installed during the construction of a building than at a subsequent period. It has become customary in large cities to wire all new buildings, even though no arrangement has been made to supply them immediately with electric current. In some cases interior conduits are placed in a new building with the intention of subsequently wiring the building.

It is often convenient to be able to turn

HOUSE FIXTURES AND WIRING. 267

a number of lamps on or off at some distance from them. For this purpose, besides the key frequently provided in the socket for turning on or off the individual lamp, *lamp-switches* are placed in the branches, or in the mains, whereby all the lamps supplied by said branches or mains may be lighted or extinguished at once. The size and character of the lamp switch will depend upon the current strength it is intended to control. Switches may be divided into two general classes; namely, *single-pole switches* and *double-pole switches*. In a single-pole switch, as the name indicates, the connection is broken on one side only of the two supply conductors; in a double-pole switch, it is broken on both sides. Switches are made in a great variety of forms, a few of which are illustrated in the accompanying figures. Fig. 84, shows a form of simple single-pole

268   ELECTRIC INCANDESCENT LIGHTING.

switch. By turning the key *K*, the spring *S*, is brought into contact with the supply spring *P*, thereby ensuring the closing of the branch circuit of the lamp through the

Fig. 84.—Simple Form of Single-Pole Switch.

switch terminals *A* and *B*. Single-pole switches should never be employed, except for a small number of lamps. Fig. 85, shows a different type of switch. *S*, being a single-pole switch, with two terminals, and *D*, a double-pole switch with four.

HOUSE FIXTURES AND WIRING. 269

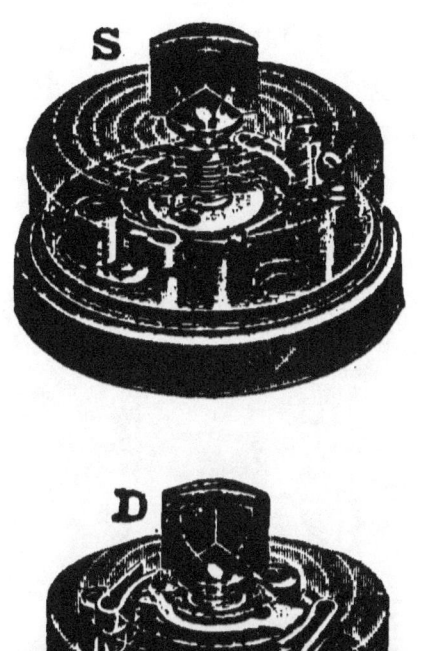

Fig. 85.—Single- and Double-Pole Switches.

In the latter case, two of these terminals are connected with the branch wires and the other two with the supply mains.

Fig. 86, shows a larger form of double-pole switch, intended for a comparatively large number of lamps, say 100. Generally, care

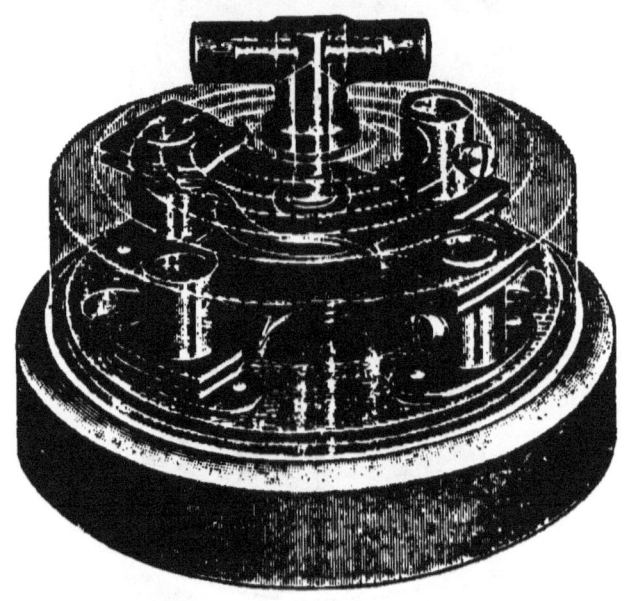

FIG. 86.—LARGER FORM OF DOUBLE-POLE SWITCH.

is given to make the contacts of considerable surface area, in order that no undue heating may arise from imperfect contact. Fig. 87, represents a different type of

Fig. 87.—Double-Pole Switch.

double-pole switch. Here the turning of the handle forces a lever into or out of a groove, whereby the contact pieces, *A B* and *C D*, are either insulated or are connected together.

Fig. 88, represents a form of switch which is sunk into the wall, so that its plane outer surface is flush with the surface of the wall. The switch is, therefore, usually called a *flush switch*.

In normal operation, the current which passes through the conductors in a building is only that which is necessary to supply the high resistance lamps connected with them. If, however, an accidental *short-circuit*, or direct connection, were to take place between the positive and negative mains in any part of the building, the supply wires connected with those mains

HOUSE FIXTURES AND WIRING.   273

would be apt to receive a rush of current that might render them white hot. In order to prevent the danger arising from this overheating, a form of *automatic*

Fig. 88.—Flush Switch.

switch has been designed, which immediately opens the overloaded circuit, and thus interrupts the current. The automatic switch invariably employed in incandescent mains is quite simple in its construction and operation. It is called a *fuse cut-out*, or *safety fuse*, and consists essentially of a strip or wire of lead, or fusible alloy, interposed in the circuit. The area of cross-section of the fuse strip is such that, while it will readily carry the normal current intended to be supplied by the conductors with which it is connected, it will immediately fuse on the passage of an abnormal current.

Fig. 89, shows a form of *branch cut-out;* i. e., a fuse block inserted between a pair of branch wires and the mains supplying them. These *branch blocks* consist of a glazed porcelain base, provided with two grooves in which are four terminals A, B, C and D, one pair of which, say A and B, are connected to the branch wires, while the other pair, C and D, are connected to the mains. A, is connected to C, and B, to D, through two strips S, S, of fusible

HOUSE FIXTURES AND WIRING. 275

alloy, clamped at the ends under smaller screws, as shown. A suitable porcelain

Fig. 89.—Branch Block.

cover is screwed over the whole, so that the fuse is completely protected. If the current passing into the branch wires becomes dangerously strong, the fuses *S, S,* are melted or *blown,* and thus automatically interrupt the branch circuit.

Various forms of fuses are employed. A common type is represented in Fig. 90. Here the fuse consists of lead-alloy wire inserted in a small glass plug *P*, somewhat resembling the socket of a lamp. This plug screws into receptacles in the cut-out, in such a manner that when a fuse is melted or blown, it is only necessary to remove the plug and screw in a new one. The forms of cut-out shown are designed for use in connection with *two-wire* or *three-wire mains.*

A complete system of house wiring includes a number of fuses. Large fuses are inserted in the service wires, where they enter the cellar, these being usually

HOUSE FIXTURES AND WIRING. 277

Fig. 90.—Plug Cut-Outs.

called the *main fuses*. The main cut-out fuses are usually placed close to a *main-switch*, where the current can be turned on or off from the house at will. All connections between risers and mains, or between mains and sub-mains, or sub-mains and branches, are usually provided with fuse cut-outs of the size corresponding to the current which they have to carry. In the circuits of the various fixtures small fuses are frequently inserted. Fig. 91, represents a few types of such *fixture cut-outs*, as they are called, shaped to conform with the fixtures for which they are intended. If a short circuit should take place, close to an individual lamp, the small fuses in the lamp circuit would melt, either at the fixture cut-out, or at the branch cut-out supplying the same. If a short circuit take place in a sub-main, the fuse at the junction between the main and sub-main would likewise melt, while finally, if a short circuit should occur in one of the larger mains, the fuses in the

HOUSE FIXTURES AND WIRING. 279

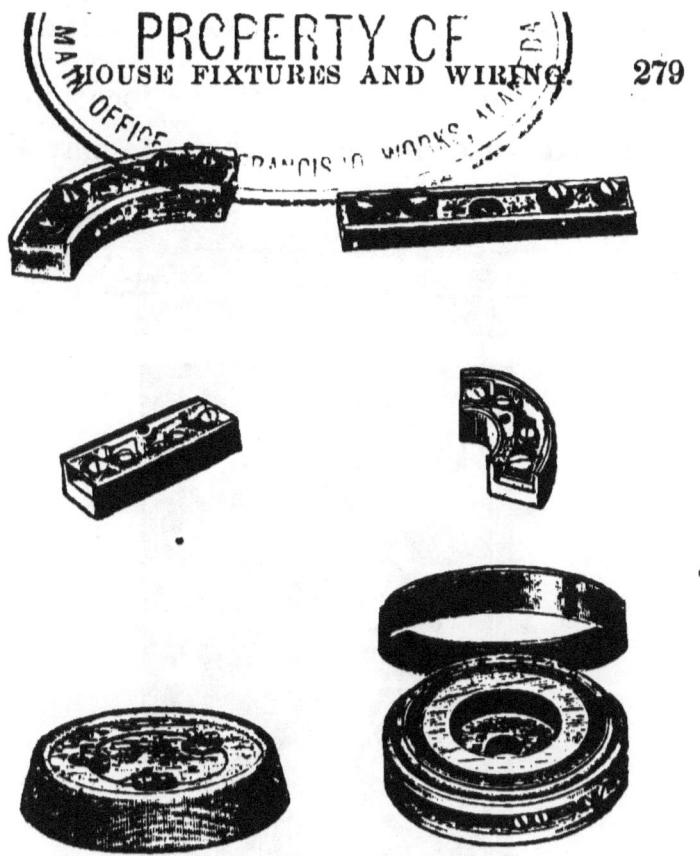

Fig. 91.—Fixture Cut-Outs.

main cut-out, where the service wires enter the building, would, probably, be instantly blown.

In all large installations, it is important

to be able to detect quickly the location of a melted fuse in order that it may be

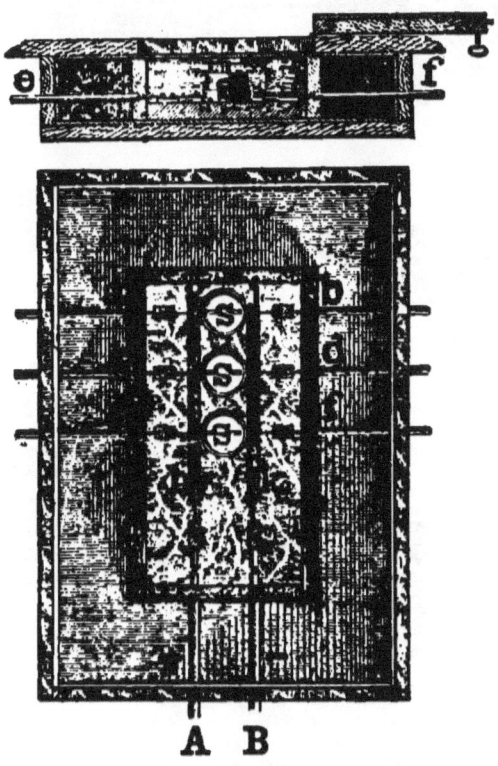

Fig. 92.—Distribution Box.

replaced and the extinguished lamps restored as soon as possible. This is fre-

HOUSE FIXTURES AND WIRING. 281

quently done by collecting all the fuses belonging to some particular portion of the system at a point called a *distributing*

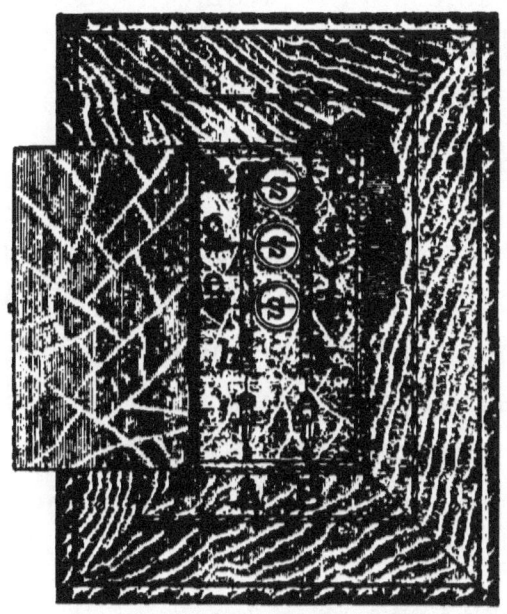

Fig. 92a.—Distribution Box.

point, usually at a junction of risers and mains, or mains and sub-mains. The *intake wires; i. e.,* those which feed the box, are usually brought to a pair of metal bars in a box called a *distribution box* lined with some fire-proof material let into the wall. The *out-put wires;* i. e., those which take their supply from the box, are attached to separate terminals which maintain connection with the metal strips, through safety fuses of the proper size.

Figs. 92 and 92A, illustrate a particular form of distribution box, which is let flush into the wall and lined with fire-proof material. AB, are the two intake wires. This box is intended for use in connection with a two-wire system. A, is placed in electric connection with the metal strip gh, and B, with the metal strip kl, through two safety fuses. The out-put wires are ab, cd and ef; a, c and e, being positive and b, d and f, negative. These wires are all placed in electric connection with their respective strips, each being provided with a safety fuse, as shown. *s, s, s,* are three switches, for controlling their independent circuits. Fig. 92A shows the box with its cover in position, but with the door opened. At the top of Fig. 92 is a cross-section of the box. In many cases these boxes have glass doors through which the condition of the fuses can be readily inspected.

CHAPTER XIII.

STREET MAINS.

In small towns, systems of incandescent distribution are effected by means of *overhead wires* or *overhead main conductors;* that is to say, both feeders and mains are insulated wires supported on poles. This method of construction is adopted, both on the score of economy and the ease of inspecting, repairing and connecting the wires. In large cities, however, where the number of such conductors is necessarily greatly increased, and where, moreover, multitudinous conductors are required for other than electric-lighting systems, the wires are, to a greater or less degree, necessarily buried underground.

Three methods are applicable to underground conductors; namely, subways, conduits and tubes. Of these methods, the first two provide means whereby the wires can be replaced or removed with greater or less readiness. By the third method, the tubes are actually buried in the ground and need excavation for examination or repair.

A *subway* differs from a conduit, in that it consists of an underground tunnel of sufficient dimensions to permit the passage of a man. Underground subways, unquestionably provide the readiest means for operating an extended system of conductors. There are, however, two serious difficulties that lie in the way of their extensive adoption; namely, expense, and want of room. In our larger cities, an unfortunate lack of uniformity has existed in the mode of use of the space beneath the streets, and pavements, for the location of the sewer, gas and water-pipes, steam heating pipes, and the various systems of electric conductors which are to-day so imperatively needed in a modern city. Unfortunately, in too many cases, the location of these has been placed under the control of different and frequently antagonistic officials. A lack of space has consequently resulted, so that in most of our large cities, the construction of a subway system would require an entire reconstruction of the systems of sewers, water, gas and electric mains.

Where a subway is employed, it is necessary to ensure its complete drainage and also to provide for an efficient system of ventilation, whereby the accidental leakage of gas into the subway shall not produce explosive mixtures with air.

The *conduit* affords a much readier and more easily applied system for underground mains or conductors. A conduit differs from the subway in that it merely provides a space for the wire or cable. Various forms of conduits have been devised, but all consist essentially of means whereby tubes, intended for the reception of cables, and generally of some insulating material, are buried in the ground. They are provided with *manholes*, or a free space in the street, extending below the level of the conduits, large enough to admit a workman. When it is desired to introduce a wire into a conduit, or to replace an injured wire, the wires are drawn into or from the conduits at the manholes.

Figs. 93 and 94, illustrate in section a conduit formed of creosoted wood, laid together in sections, so as to leave cylindrical spaces between them, through which the wires or cables may be drawn. This system is frequently used for the

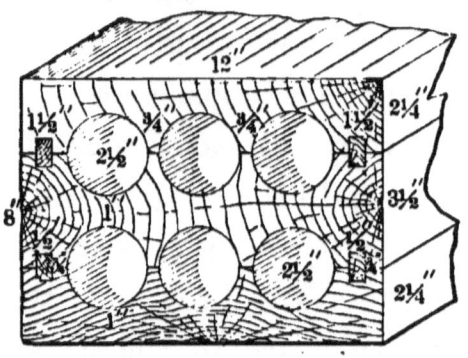

FIG. 93.—CONDUIT OF CREOSOTED WOOD.

reception of lead-covered high-tension wires, and telephone cables.

While overhead conductors may be unobjectionable in small towns or villages, yet, in large cities, where the need for wires is great and, moreover, is constantly increasing, a condition of affairs might readily be brought about such as is shown in Fig. 95, which represents the condition of a street with the

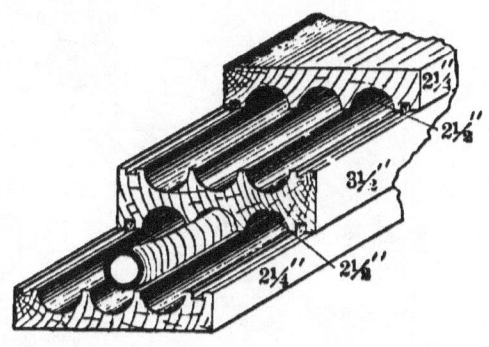

Fig. 94.—Conduit of Creosoted Wood Cut Away to Show Structure.

many aerial wires that are to be expected. Contrast this with the altered appearance of the same street, when the wires were placed in underground conduits as shown in Fig. 96, and the advantage of underground wires, from an æsthetic standpoint, is manifest.

290 ELECTRIC INCANDESCENT LIGHTING.

FIG. 95.—VIEW OF CITY STREET WITH OVERHEAD WIRES.

STREET MAINS.

Fig. 96.—View of City Street after Removal of Overhead Wires.

Fig. 97, represents the third method for underground conductors, viz.; the *underground tube*. Here an iron pipe is employed, containing three insulated conductors, and intended for use in a three-wire

FIG. 97.—TUBE CONTAINING THREE SEPARATELY INSULATED CONDUCTORS.

system of distribution. The iron pipe is provided for the purpose of protecting the conductors from mechanical injury. Fig. 98, shows cross-sections of different sizes of these tubes. A, B and C, are the cross-sections of the copper conductors, surrounded and supported by a bituminous insulating material. The outer ring is the section of the iron pipe.

As the above form of underground tube is to-day in extended use, a description of

STREET MAINS. 293

its manufacture will not be out of place. The iron pipes are made up in lengths of twenty feet. The copper conductors, cut off to the right length, are prepared for

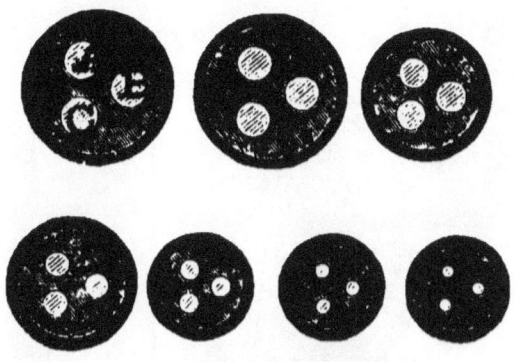

Fig. 98.—Main Tubes.

placing in the tube by wrapping each with a loose or open spiral of rope. Three rods are then assembled and held together, in the position shown in the cross-section, by wrapping them with a tight wrapping of rope. The rods so assembled, are now placed in a length of tube, and one end of the tube is closed

294 ELECTRIC INCANDESCENT LIGHTING.

with a plug. The tube is then filled with hot bituminous insulating material. The tube is finally closed by a second block. There will thus be provided,

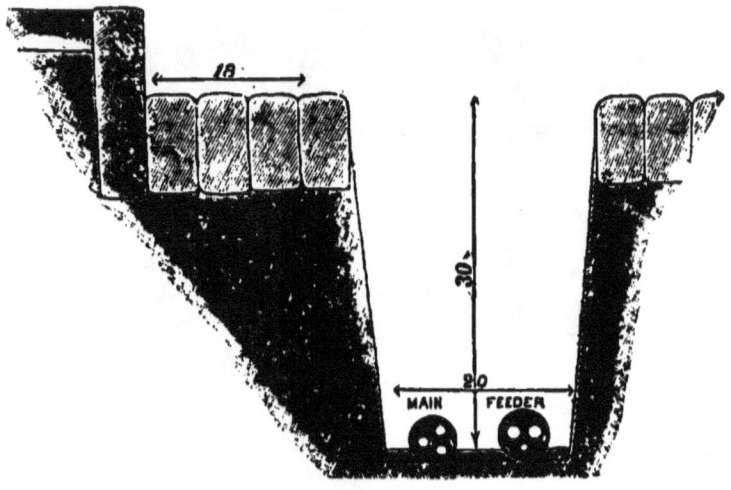

FIG. 99.—SECTION OF STREET TRENCH CONTAINING A FEEDER AND MAIN TUBE.

lengths of pipe twenty feet long, containing three insulated conductors with their extremities projecting at the ends.

The underground tubes so formed are

sent out from the factory in the twenty-foot sections described. They are subsequently connected to one another, while in position in the underground trench prepared to receive them. Fig. 99, shows a section through a trench provided for a feeder and a main tube. The trench is usually 30" deep, as shown, and is situated in the street a short distance from the curb, the pipes being laid end to end at the bottom of the trench.

It now remains to connect the separate lengths of the underground tubes. For this purpose *coupling boxes* are provided as shown in Fig. 100. *A*, shows a coupling box suitable for a street connection, and *B*, a coupling box for a right-angled connection. The coupling box is formed of a cast iron shell made in halves, connected together by bolts passing through flanges, and clamped over collars secured at the ends of the tubes. The

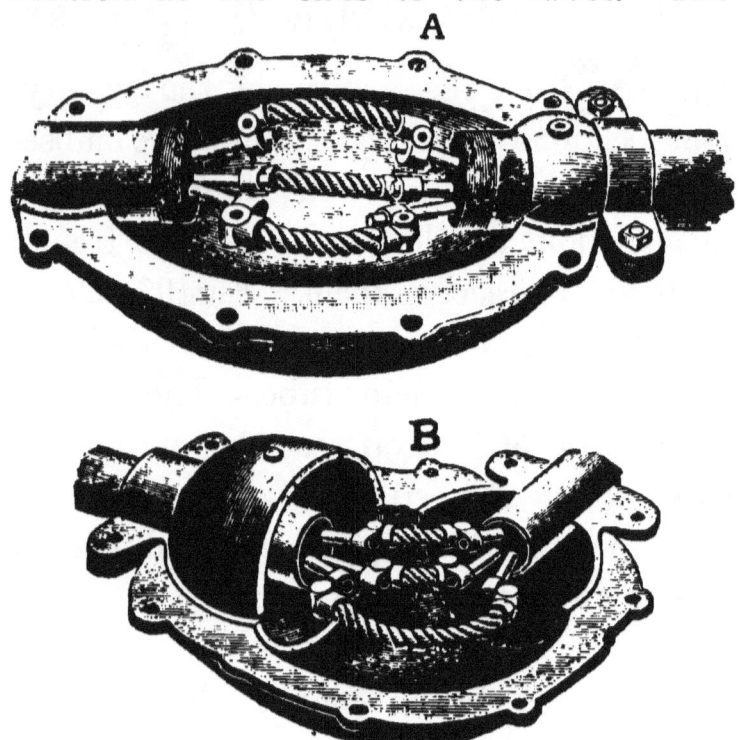

Fig. 100.—Coupling Boxes.

collars are first secured in place. The lower half of the box is then fitted and the three flexible copper stranded connectors are forced on the ends of the conductors as shown. By the aid of a torch, the joints are all heated to a sufficiently high temperature, and solder is melted into them, thus forming a good metallic junction. The upper half of the box is then fitted into place, the two parts clamped together by the bolts, and melted bituminous compound is poured into the box through an aperture in the top. The coupling box is then closed as shown at J in Fig. 97.

Fig. 101, shows a form of *branch coupling box* suitable for a house-service connection with the mains. In this case, A and B, are *main tubes* passing along the street, while C, is the house service tube. Here the coupling box is shaped to conform to the requirements of the triple connection as shown.

298   ELECTRIC INCANDESCENT LIGHTING.

Fig. 102, shows cross-sections of *feeder tubes*. In these tubes, the neutral conductors are of smaller size than the two outside conductors, since the system is

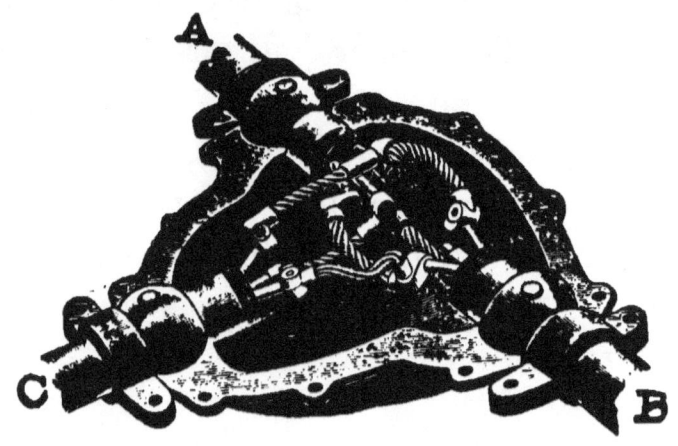

Fig. 101.—Branch Coupling Box.

arranged to be nearly balanced as to load with reference to the neutral. The three small black wires shown, are called *pressure wires;* i. e., small insulated copper conductors which are returned from the *feeding point*, or point of junction between

STREET MAINS. 299

feeder and mains, to the central station, so as to indicate in the central station, the pressure which is supplied to the mains. Thus, if the drop in the feeders, be say 15

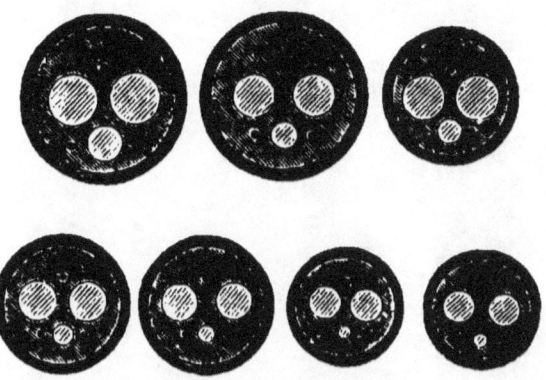

Fig. 102.—Feeder Tubes.

volts, and the pressure at the bus-bars, in the central station, be 130 volts on each side of the system, or 260 volts across outside bus-bars, the pressure in the mains at the feeding point will be 115 volts on each side, and this will be the pressure carried by the small pressure wires back to the indicators at the central station.

300   ELECTRIC INCANDESCENT LIGHTING.

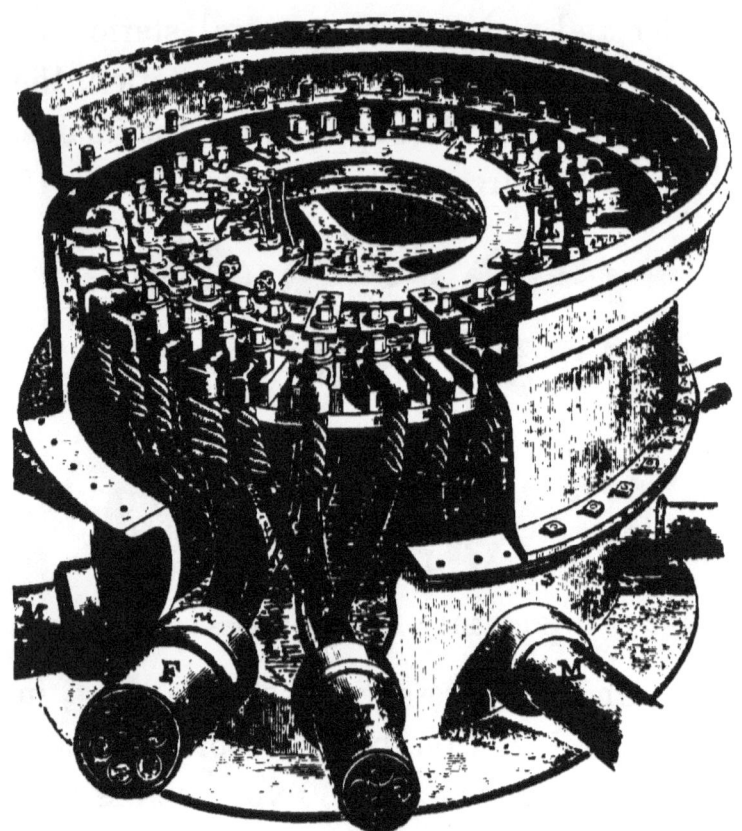

FIG. 103.—JUNCTION BOX.

Fig. 103, shows the interior and Fig. 104, the exterior of a *junction box;* i. e., a box situated at the junction of two streets, or

STREET MAINS. 301

at a feeding point where a feeder joins the mains. This box is of cast iron, and is buried with its surface flush with the street level. The tubes enter the sides of

FIG. 104.—JUNCTION BOX.

the box at the lower level. The conductors are connected by flexible connections with brass pieces supported on insulating rings. These pieces are marked + and − according to the polarity of the conductors connected with them.

Three metallic rings, insulated from each other, are provided in the box corresponding to the three conductors in the tubes. All the positive conductors are connected across to the positive ring, through fuse strips; all the negative conductors are connected to the negative ring; and all the neutral conductors, to the neutral ring. In this way all the positive conductors are placed in electrical connection with each other, and also the negative and neutral conductors. The box is made water-tight by lowering a cover over the bolts seen in the ring in the upper part, screwing down nuts upon these bolts, and pouring in bituminous compound in a melted state over the bolts. F, is a feeder tube containing the pressure wires and five conductors, a pair of negatives, a pair of positives, and a neutral. MM, are main tubes; p, is a slab or strip of insulating material, supporting the connections for the pressure wires, which are connected through fuses with the three rings in the box. Fig. 104, shows the appearance of the junction box when closed with an ornamental cover.

CHAPTER XIV.

CENTRAL STATIONS.

IF we could follow the buried conductors through the streets, up-stream, that is, toward the supply, we would finally reach a building toward which all these wires converge. This building constitutes what is called a *central station*. In it we will find the means for generating the current which is sent through the street mains and feeder wires, through the service conductors into the house, and finally through the risers and house mains and branches, to the lamps. Limiting our description of the station to one in which *continuous electric currents* only are generated; *i. e.*, currents which always flow in the same direction, and which are suitable for use in the three-wire system of buried conductors just described, and supposing, as is generally the case, that steam power is employed, we will find that the apparatus can be readily grouped into three general classes; namely,

(1) The dynamos.
(2) The engines.
(3) The boilers.

Directing our attention in the station first to that part of the building at which the feeders enter from the street, we will find them connected to a device called a *switchboard*. This consists essentially of a fire-proof frame supporting a number of metallic terminals provided for connection with the feeders and also with connections designed to receive conductors from the generators. A number of instruments are mounted on this switchboard, consisting of *ammeters* to show the strength of current, and *voltmeters* to show the pressure on the various feeders, while switches are provided for opening and closing the various circuits.

Fig. 105, shows a partial view of a switchboard in a central station. *S, S, S,* are rows of massive switches mounted upon fire-proof slabs of insulating material. Above the switches are the voltmeters and ammeters, while at *F*, are the *field regulating boxes* for controlling the pressure of the generators. During the daytime, the load on the mains is principally due to motors and is considerably less than the night load. As evening approaches, the load increases, as is shown on the ammeters at the station. As soon as it becomes

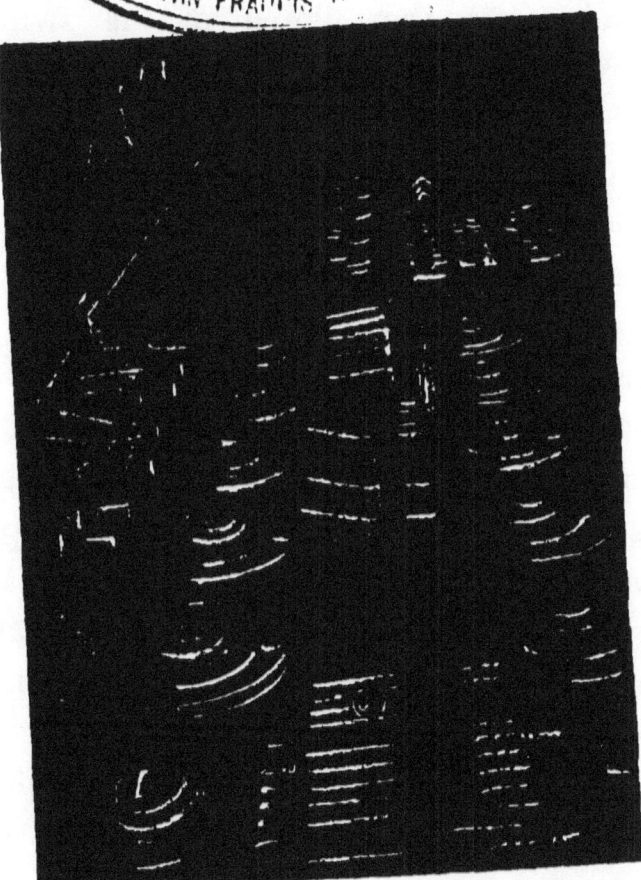

Fig. 105.—Partial View of Switchboard

necessary to introduce another pair of dynamos, the engine driving them is started, the pressure of the dynamos is brought up to that required, and the switches are then closed at the switchboard, thus connecting the new generators with the feeders. The reverse process is adopted as the load diminishes, and it is unnecessary to any longer maintain the extra generators in the circuit.

Fig. 106, shows a type of *generator unit* frequently met with in large central stations, for low-pressure incandescent-light distribution. This figure represents a portion of the engine room in a large central lighting station, and shows two generator units at *A* and *B*, respectively. *A*, is a vertical condensing, triple-expansion engine, *E, E, E,* whose main horizontal shaft drives at each end the armature of a dy-

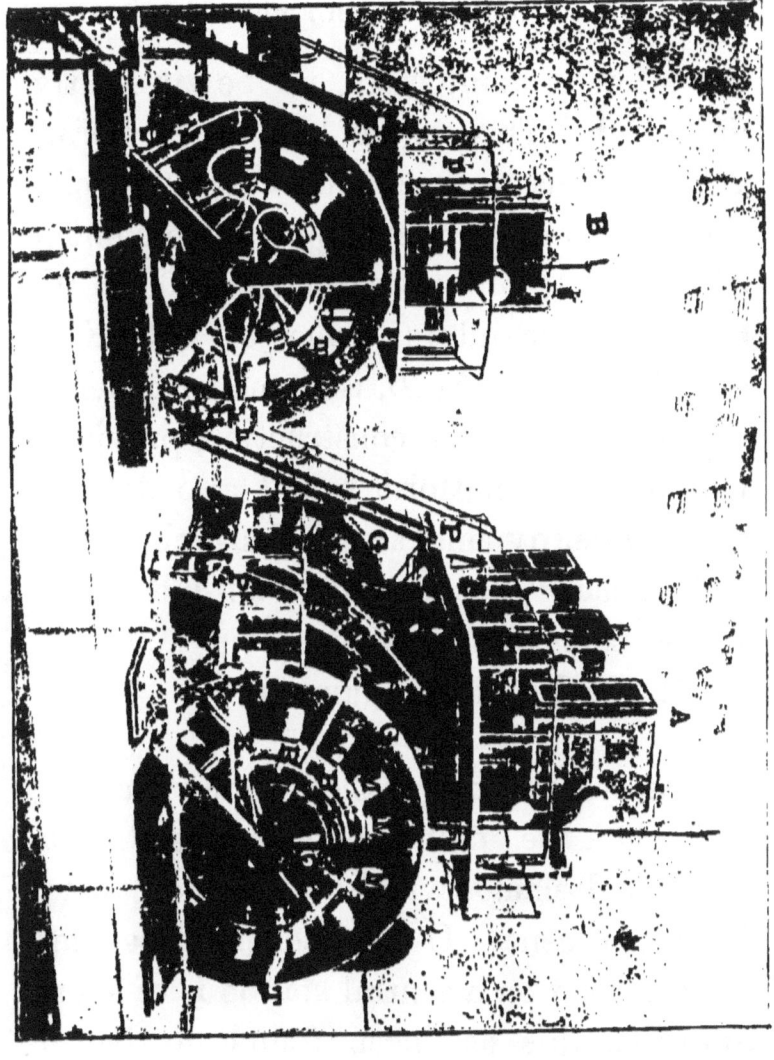

Fig. 106.—Central Station Generator Units.

namo or generator G, to be presently described. This engine has two platforms P, P. This engine runs at 120 revolutions per minute, developing a maximum of 800 HP; or, approximately, 600 KW, the two generators being of 200 KW capacity each. At B, is a smaller generating unit of similar construction, provided with a single platform p, and also driving two smaller dynamo armatures, each of 100 KW capacity, the maximum out-put of the engine being 300 KW, or about 400 HP, at 172 revolutions per minute.

It is evident, under these circumstances, that when the engine is developing its full load, the dynamos will be overloaded about forty per cent. Owing to the fact that the full load on a central station is of short duration, this has been found to be an economical practice.

In order to provide power to drive the engines and dynamos a *battery of boilers* is installed. Boilers are of various types. Since an incandescent lamp requires an activity of 50 watts, or about $\frac{1}{15}$th of one horse-power at its terminals, and since losses occur in the feeders, mains and house wires of, perhaps, ten per cent. on an average, the activity per lamp at the dynamo terminals, in a central station, will be about 55.5 watts. Moreover, since an average loss of say fourteen per cent. occurs in the engine and dynamo, the activity per lamp, generated by the engine, must be about 64.5 watts, or about the $\frac{1}{11.5}$th of a horse-power, so that on an average, one indicated horse-power at the engine, represents 11.5 sixteen-candle power incandescent lamps of 50 watts each. In other words, the average commercial efficiency of such a system of distribution, starting from the indicated horse-power of the engine, is, approximately, 77.5 per cent. About 17 pounds of steam, at 160 pounds pressure, are required per indicated horse-power-hour with engines of the type shown. This represents about 2.9 pounds of coal per horse-power-hour delivered electrically in consumers' lamps.

A large central station may readily supply 60,000 incandescent lamps or more. Assuming that 60,000 is supplied at maximum load, the boiler power needed will be correspondingly great. Take, for example, the central station that was required to supply the buildings and grounds of the World's Columbian Exhibition at Chicago, in 1893, with electric light. There were employed for lighting this exhibition,

CENTRAL STATIONS. 313

about 100,000 incandescent lamps and about 5,000 arc lamps. There was naturally required for this purpose, as well as for driving the machinery in the buildings, a very great amount of power.

Fig. 107, shows a view of the main boiler plant in the above exhibition. Here batteries of boilers, representing an aggregate of 24,000 horse-power are shown. D, D, are the main doors, d, d, the fire doors, and beneath these latter the ash-pit doors. G, is the steam gauge to show the steam pressure in the boiler above that of the atmosphere; g, the water gauge; M, the steam drum, and P, the main steam pipe.

To the student of the incandescent lamp, the most important features of the central station are the *dynamos* or *generators* designed to supply the current to the lamps. Limiting our attention now to continuous-current dynamos, we will find these to be of a variety of types, although all operate on essentially the same principle.

Broadly speaking a dynamo-electric machine or generator, is a device whereby electromotive forces, and from these, electric currents are produced, generally by the revolution of conductors through magnetic flux. The magnetic flux is produced by *field-magnet coils*. E. M. Fs. are set up in the coils on the revolving portion, called the *armature*. These E. M. Fs. are generated in the armature coils in successively opposite directions, as they pass each pole; consequently, they need to be commuted, or caused to assume the same direction in the external circuit. This is effected by a device called a *commutator*.

Fig. 107.—24,000 Horse-Power Boiler Plant at the World's Fair.

Thus Fig. 106, represents *multipolar generators*, there being fourteen magnets or poles M, M, M, on the large dynamo, and eight magnets or poles, on the small dynamo. These magnets form part of the massive stationary iron frames $F'\,F\,F, f\,f\,f$, supporting the outboard bearings O, o. In the cylindrical bore of these field-magnet poles, run the armatures, which are rigidly secured to the main engine shaft. A light metallic frame supports a number of pairs of *brushes* B, B, upon the surface of the commutator, there being as many pairs of brushes as poles, so that A, has fourteen pairs of brushes upon its commutator and B, has eight pairs. These pairs of brushes, or *double brushes*, as they might be called, are insulated from their supporting frame, but are connected in alternate pairs with the main terminals T, T'. Thus the first, third, fifth, etc., brushes are connected to one terminal, and the second, fourth, sixth, etc., with the other. *H, h,* are handles for rocking the entire brush-holder frame to-and-fro within a small angular range around the axis, so as to carry all the brushes forward or backward upon the surface of the commutator. This adjustment is made to prevent *sparking* at the brushes from the current which they carry at different loads. The main terminals *T, T,* are connected to switches on the main switchboard of the station.

Fig. 108, shows in greater detail another form of central station generator of the same type. *F, F, F,* is the field frame. *M, M, M,* are the field coils, six in number, consisting of coils of insulated wire wound upon soft steel cores bolted radially to the field frame, as shown. *A,* is the armature whose surface runs within the cylindrical polar space formed by the poles of the field magnets. $C, C,$ is a commu-

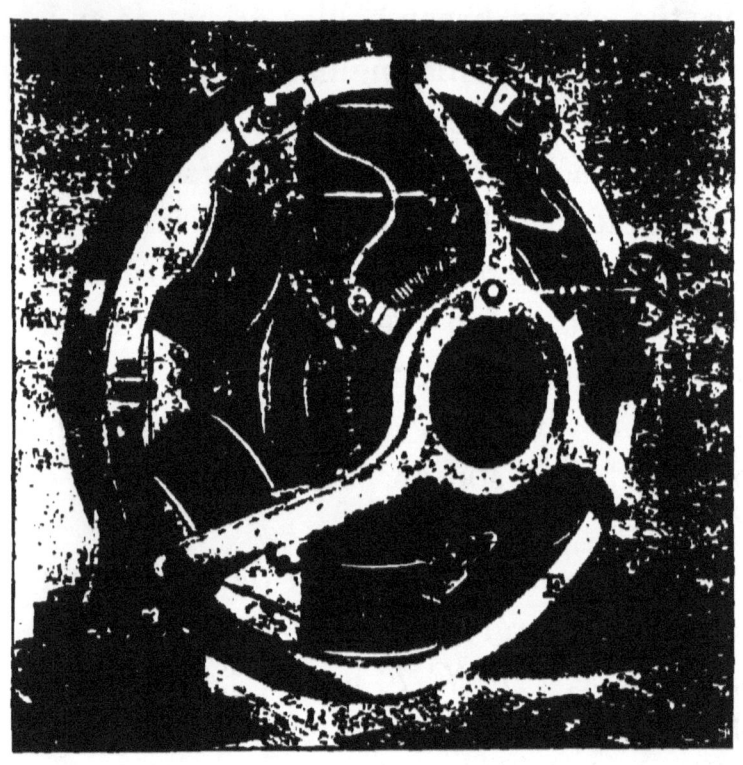

Fig. 108.—Six-Pole Generator.

tator, consisting of copper strips, rigidly secured in a cylindrical frame, and insulated conducting loops wound upon the armature. *B, B,* are the brushes, of which there are six pairs, connected to two metallic rings, one ring being in connection with three alternate pairs. These rings are finally connected to the main terminals *T, T,* by cables, as shown. The current supplied by the armature is collected by each brush, leaving the armature at the three positive brushes, and, after traversing the external circuit of the feeders, mains, house wires and lamps, returning to the armature through the three negative brushes, thereby completing the electric circuit. *H,* is a handle for advancing or retreating the brushes over the surface of the commutator. The machine shown has 50 KW capacity, that is to say, it will deliver at its terminals activity to the amount of 50 KW, (400 amperes at a pressure of 125 volts, or 400 × 125 = 50,000 watts). It requires about 75 horse-power or about 56 KW to drive it at full load. Its gross weight is 6,500 pounds, representing an output of, approximately, 7.7 watts per pound of total weight. It has a shaft five inches in diameter, and occupies a space of 30" × 61" × 51" in height.

The armature of the preceding machine is illustrated in greater detail in Fig. 109. It consists of a metal cylinder, which carries on its outer surface a laminated iron core, built up of annular sheet-iron discs, clamped side by side, so as to form, when assembled, an almost complete cylindrical external surface of iron, $A\ A$. Gaps are left, at suitable intervals between adjacent sets of discs, to provide for the ventilation of the armature, by means of the passage of air by centrifugal force from within outward, to aid in the cooling of the ar-

Fig. 109.—Armature of Generator.

mature, when operated. The outer surfaces of the iron discs are provided with longitudinal slots, parallel to the axis of rotation. These slots are designed to receive the armature conductors. One hundred and sixty-eight of these slots are thus provided to receive one hundred and sixty-eight sets of conductors.

The winding adopted in the armature is conducted as follows: At A, a pair of rods or wires comes through a slot across the armature surface. A pair of flexible copper strips, insulated from their neighbors, are soldered to the rods at a, and run down to b. These connect two adjacent commutator bars at b. From this they run back behind the connections W, W, to the slot at c, and then across the armature surface underneath the two rods or wires which are seen to approach at c. Having reached the opposite side of the armature at c', they descend obliquely on the other face to a point opposite *d*, and then ascend to the armature surface at a point *c'*. Here the rods connected therewith return across the armature surface to the point *e*, descending again to the commutator at the point *f*. In this manner a pair of conductors zig-zag across the armature, via *g i k*, emerging again one slot in advance of *a*, and so on. In this manner it must always happen that the wires in the slot *a*, pass under one pole, the wires *c e g i k* will also be passing under the other poles. The effect of the commutator is to enable the brushes to collect all the currents which are being generated in the various wires passing under the different poles and to unite them in the external circuit.

CHAPTER XV.

ISOLATED PLANTS.

For the general purposes of house lighting, where comparatively few lamps are required, it is more economical for the householder to rent his electric power from a central station, than it would be for him to erect and maintain his own plant. Since a plant requires boilers and engines, it would follow, unless a fairly considerable number of lamps is required, that the extra expense of the installation as well as the services of an engineer, or engineer and fireman, would make it much cheaper in such cases to take the service from the street mains as supplied by the nearest central station. There are many cases, however, in which either the number of lights, or the circumstances are such, as to warrant, in point of economy, the maintenance of what may be called an *isolated lighting plant*, in contradistinction to a central-station lighting plant.

It is clear that when the number of lights required reaches a certain limit, it may be preferable to maintain an isolated plant, rather than to rent the light, since a large building or plant would thus possess the advantage of being independent of the running of the central station, or of accidents which might occur to the street mains. Moreover, a large isolated plant, being necessarily circumscribed in the area of its distribution, would require a much smaller outlay or expenditure in copper, and, consequently, may be built to produce light cheaper than a central station.

But even in cases where the number of lights is not very great, circumstances may arise where it would still be economical to establish an isolated plant. Such cases would be found, for example, in manufacturing establishments, where boilers and steam engines have necessarily to be maintained in action during the time that light is required. Again, isolated plants are necessary in sections of country remote from central stations. A brief description of isolated plants for the supply of incandescent lighting will, therefore, be of interest.

Any suitable form of dynamo can be used for an isolated plant. The dynamo may be driven either *directly* from the engine shaft or by means of *belting*. The

ISOLATED PLANTS. 327

direct connection requires less floor space, is somewhat more efficient, and saves wear

FIG. 110.—BELT-DRIVEN GENERATOR FOR ISOLATED PLANTS.

FIG. 111.—DIRECT-DRIVEN GENERATOR FOR ISOLATED LIGHTING.

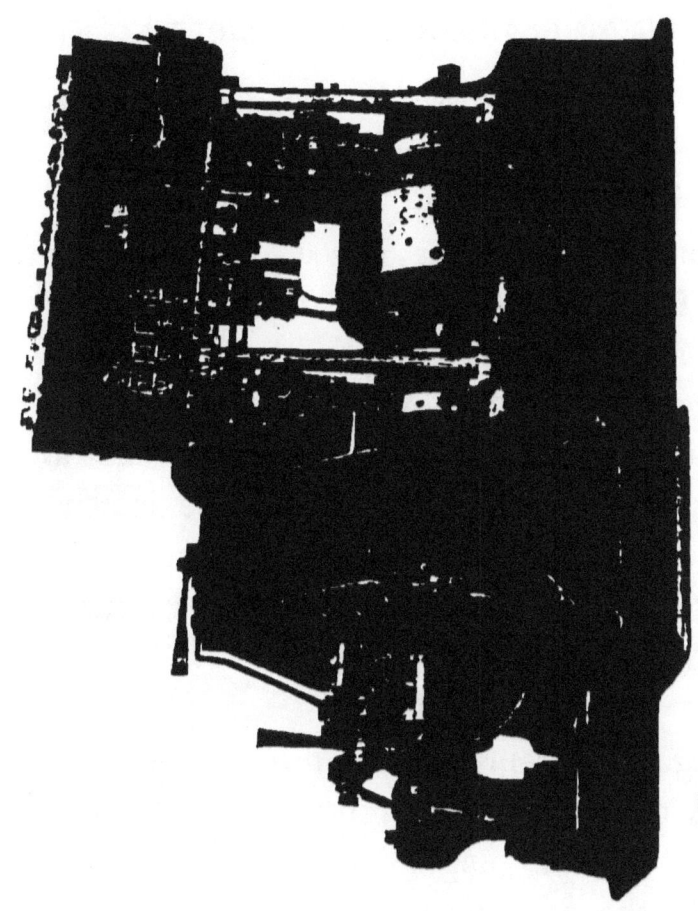

Fig. 112.—Direct-Driven Quadripolar Generator.

of belting, but possesses the disadvantage of requiring the engine to run at a comparatively high speed, and the dynamo at a comparatively low speed. This means that for all sizes of generator below 50 KW at least, the cost of a direct-driven plant is greater than that of a belt-driven plant, because a slow-speed dynamo means a heavier and, consequently, a more expensive dynamo, and a high-speed engine is more difficult to maintain in running order.

Fig. 110, represents a *bipolar*, or two-pole, generator, driven by a belt from a small vertical engine with its governor inside the fly wheel.

Fig. 111, represents a *quadripolar* or four-pole direct-driven generator, suitable for isolated plants. In this case the en-

Fig. 113.—The Three-Wire Isolated Plant.

gine is completely enclosed in a cast-iron shell and runs in oil.

Fig. 112, represents a type of isolated lighting plant employed on board ships and supplied to several vessels in the United States Navy. This quadripolar generator delivers 200 amperes at 80 volts pressure, or 16 KW, at a speed of 400 revolutions per minute. It is directly connected as shown to the vertical marine engine.

Fig. 113, represents an isolated three-wire plant, consisting of a 100-horse-power horizontal steam engine, driving two 32 KW quadripolar generators at a speed of 270 revolutions per minute. The magnet poles in this case are inside the armature and the outer surface of the armature has its insulated conductors bare, so as to form a commutator, upon which the sets of

ISOLATED PLANTS. 333

brushes shown in the figure can rest. The switchboard for controlling the various circuits is represented behind the machine on the right hand side of the figure. The apparatus required for such a switchboard is similar in kind to that of a central station, but usually is smaller and less complex.

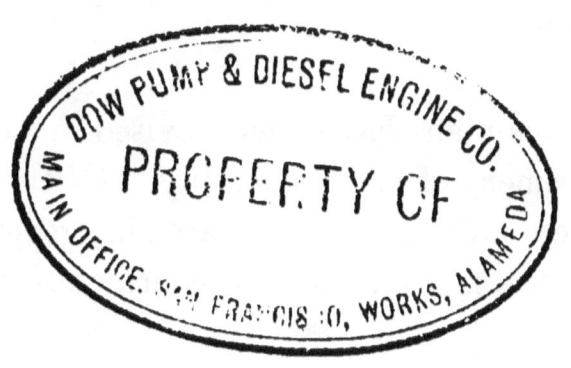

CHAPTER XVI.

METERS.

The sale of any product requires some suitable unit of measure. Since electric power is undoubtedly a product requiring, as it does, the establishment of an expensive plant for its production, a unit of measure is necessary, as well as an apparatus whereby the number of units delivered to the customer by the producer may be measured.

Various plans have been devised for the measurement of electric supply. Of these, however, only two forms are in general use for the measurement of continuous-cur-

METERS. 335

rent supply. One of these forms measures the quantity of electricity delivered to the consumer in units of supply called *ampere-hours*, while the other measures the energy in *watt-hours*. Each of these forms of apparatus is called a *meter*.

Fig. 114, shows a form of *electrolytic meter* indicating the supply in ampere-hours. This apparatus depends for its operation on the fact that an electric current, when sent through a solution of zinc sulphate, will effect a decomposition of the solution, depositing metallic zinc on a zinc plate connected with the negative terminal, and dissolving or removing an equal quantity of zinc, from a plate of zinc connected with the positive terminal. The indications of this meter are obtained by carefully weighing the plates, before and after the supply they are to measure has passed through them. After the meter has been in use for some time, it will be found that the zinc plate connected to the positive terminal has decreased in weight,

FIG. 114.—ELECTROLYTIC METER.

and that connected with the negative terminal has increased in weight. This difference of weight is a measure of the number of ampere-hours that have passed through the cell.

In the interior of the meter box, in the upper part, are two large strips of german silver R, R, carrying the current to be supplied to the lamps, and offering a definite small resistance to its passage. The positive and negative supply wires enter the box at P and N, and are secured to the terminals P and N, inside. As the meter shown is a three-wire meter, it consists of two meters, in one box, one being for the supply on the positive main, and the other for the supply on the negative main. If the number of lamps lighted in a house, on each side of the system, were always the same, the records of these two meters would be equal. When the current passes through the resistance $R\,R$, it establishes a certain drop of pressure, as already explained in Chapter III. This drop amounts to about 0.4 volt at full load. In a pair of derived or shunt circuits, connected across the extremities, of each strip of resistance R, is placed a pair of small glass bottles with resistances wound on a spool s, behind them. The total resistance of the bottle, or plating bath, and the spool in its circuit, bears such a relation to the resistance of the shunt R, that each milligramme of zinc electroplated on the surface of the negative zinc plate z, represents a definite number of ampere-hours of current supplied through the meter according to its size. The plates z, z, are made of zinc and mercury alloy, and are separated from each other by hard rubber washers. Copper rods extend upwards from these plates through the corks c, c, to the copper clips p, p. There are two bottles for each side of the meter, so that a duplicate record is kept on each side of the supply. It is usual to renew the bottles once a month. The bottles after being removed are emptied, the zinc plates washed and dried, and weighed in a chemical balance, and the amount of the supply determined from the difference in weight during the month. The solution employed is of pure zinc sulphate in water, having a density of 1.11 at 60° F. The advantage of this meter is its simplicity, and the fact that all its essential working parts are removed and replaced once a month. The disadvantage of the meter, is that it does not show directly to the consumer, the amount of supply which has been delivered.

When such meters are placed in exposed situations, where the solutions might be liable to freeze, a *thermostat* is employed to maintain the temperature of the bottles above the freezing point. Such a thermostat is shown in Fig. 115. Here a metallic strip, composed of two unequally expansible metals riveted together, is so arranged, that, when exposed to a suffi-

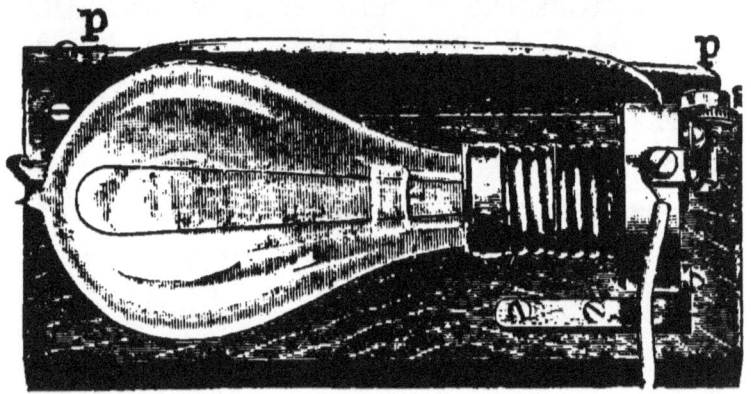

FIG. 115.—THERMOSTAT FOR ELECTROLYTIC METERS.

ciently low temperature, the unequal contraction of the metals will cause a bending or warping, which will close the circuit of a lamp through the set screw s. The lamp will then burn until the temperature rises sufficiently to allow the strip $p\,p$, to straighten and break the circuit.

METERS. 341

Another form of meter in general use is shown in Fig. 116. In this form of instrument, the number of watt-hours delivered to the consumer is registered. It

FIG. 116.—INTERIOR OF RECORDING WATTMETER.

consists of a small motor, the field magnet coils M, M, of which are in the direct circuit of supply. The armature A, placed within the field coils, revolves upon the vertical shaft S, S, with a speed proportional to the activity delivered; *i. e.*, to the product of the volts and amperes. For example, if the pressure at the house mains be 110 volts, and the current be 2 amperes, then the activity delivered would be $110 \times 2 = 220$ watts; and, in one hour, this rate of delivery would result in a supply of 220 watt-hours, while the rotary speed of the armature shaft would be proportional to this value 220. A small commutator is seen just above the field coils with its long slender brushes running back to supports behind the apparatus. The armature is placed in a circuit of high resistance across the mains, so that it absorbs a constant small amount of activity,

METERS. 343

Fig. 117.—Recording Wattmeter.

whether the meter be running or not. The shaft engages by means of an endless screw with a pinion wheel, forming part of a train work of dial-recording mechanism, similar to that of a gas meter.

The above form of the apparatus, when completely enclosed, is seen in Fig. 117. The cover is secured to the base by a wire sealed with the leaden seal *s*. The supply and output wires pass through the meter beneath. The advantage of this meter is that it enables the customer to observe the amount of power he consumes. Its disadvantage is that it constantly absorbs a small amount of power. A meter should always be installed in a dry place, and inserted in the service wires between the street main cut-out and the risers.

CHAPTER XVII.

STORAGE BATTERIES.

If the supply of electric current required for incandescent lamps was uniform throughout the twenty-four hours, it would be easy to determine the most economical generator units of boiler, engine and dynamo, in order to meet this requirement economically. Unfortunately, however, in nearly all cases the variations in the load are very great. A few hours of the twenty-four require a load greatly in excess of the average. If, in order to meet this load economically, the generating plant be subdivided into a number of small units, so as to permit them to be readily withdrawn and added as required, both the expense and the complexity of the generating system would be necessarily increased. If, on the other hand, a large generating unit be installed, it would have to be operated for the greater part of the twenty-four hours at a very small load, and, therefore, uneconomically.

The above difficulty is sometimes met by the employment of storage batteries, which are charged during the hours of light load, and discharged during the hours of heavy load, thereby equalizing the load on the station. Since, at the time of full load, the output is obtained both from dynamos and storage batteries, it is evident that a much smaller generating plant of boilers, engines and dynamos is rendered necessary.

In order to determine whether it would

be economical to install a storage battery, it is necessary to ascertain the *load diagram* of the station; that is, the curve which represents the output required to supply the

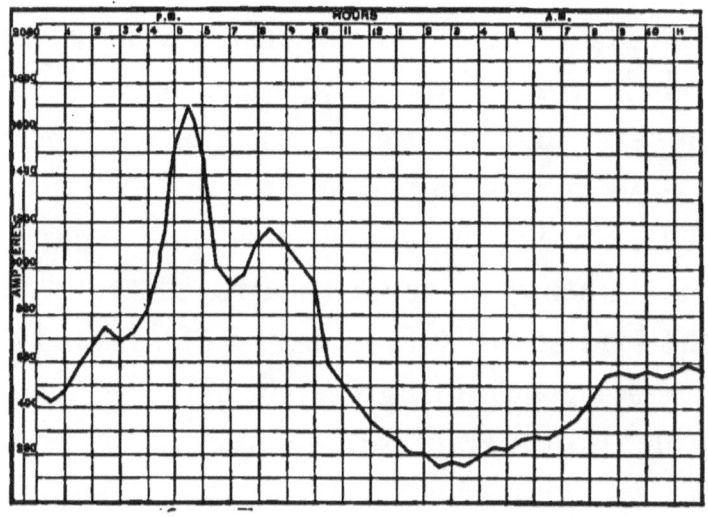

FIG. 118.—LOAD DIAGRAM.

lamps during the twenty-four hours of the day. Such a load diagram is shown in Fig. 118. In this figure the hours of the day are marked off horizontally, and the current strength delivered to the feeders is marked off vertically in amperes. An inspection of the figure will show that the *peak of the load*, that is, the maximum load of the curve, occurred at 5.30 P. M. when it exceeds 1,700 amperes, while an hour and a half earlier, or at 4 P. M., it was 800 amperes, and an hour and a half later, or at 7 P. M., it was 930 amperes. At 8.30 P. M. the load has increased, perhaps, owing to the lighting of some theatre, after which the load steadily falls until 2.30 A. M. The average load during the twenty-four hours from this diagram is 620 amperes, and since the maximum load is 1,710, the ratio of the average to the maximum is 0.363 or 36.3 per cent. This is called the *load factor*.

If the load represented in Fig. 118, were supplied without the aid of a storage battery, it would be necessary to install boilers, dynamos and engines to the extent necessary to supply a current of 1,700 amperes, although the average load is only 620. By the use of a storage battery, capable of supplying 800 amperes for, say four hours; or, a battery having a *storage capacity* of $4 \times 800 = 3,200$ ampere-hours, then the maximum load which the boilers, engines and dynamos would have to supply would be 900 amperes; or, only about half as much as in the preceding case, while the load during the daytime would be increased by the amount necessary to charge the storage batteries. Good practice requires that the charging be done during the time of the day when the load is the least; or, in this case, between the hours of 1 and 7 A. M. The smaller the load factor the greater the probability of obtaining economy in the installation of a storage battery.

Another case arises in which an advantage is derived from the use of a storage battery; namely, where the engines and boilers employed to drive the dynamo are used during the hours of darkness only, and are stopped during the day, while it is desired to maintain a few lamps during the daytime. Under these circumstances it is often more economical to establish a storage battery plant, which can be charged during the time of running at night, thus permitting the engine and dynamo to be stopped during the daytime and the storage battery to supply the few lights required.

Before proceeding to the general description of a storage battery installation, it may be well to describe briefly the principles of its operation. A storage cell differs in no respect from an ordinary voltaic cell;

for, like a voltaic cell it consists essentially of two *plates* or *elements*, called respectively the positive and the negative plate, plunged in an acid liquid or electrolyte capable of acting on one of the plates. As the result of the chemical action that occurs under these circumstances, an electric current is produced which passes in a definite direction through the electrolyte and issues from the cell at one of its terminals or poles and returns to it, after having passed through the circuit in which the cell is connected, by the other terminal or pole. In both the ordinary voltaic and the storage cell, exhaustion takes place when a certain output of electricity is yielded. In the case of the voltaic cell both the liquid, and at least one of the plates, must be renewed, while in the case of the storage cell, all that is necessary for renewal is to connect the terminals of the cell with an independent electric source, and send a current through it in the opposite direction to that of the current it yields. This current is called the *charging current*, and the cell receiving it is said to be *charged*. Since a storage cell thus derives its energy secondarily from some other electric source, it is sometimes called a *secondary cell*, in contradistinction to a primary or voltaic cell.

A great number of storage cells have been devised. Practically all that have been placed in commercial use, consist of perforated lead plates or *grids*, corresponding to the positive and negative plates of the voltaic cell, with the perforations filled, respectively, with peroxide of lead and finely divided metallic lead. These plates are associated, or placed side by side, in a solution of sulphuric acid and water. In the original form given to these cells, they consisted of a very large positive and negative plate, suitably supported at a short distance from each other, and then rolled up together in a close spiral. It was found, however, in practice, that when such a plate became damaged in one part, the entire cell had to be rejected, so that for the purpose of convenience, as well as for the ready inspection of the different parts of the plate, they are now generally made in a number of smaller plane plates, placed parallel to one another, which, when connected in parallel, are equivalent to two large plates of the same total surface.

The simplest form of such a cell would consist of a single positive and a single negative plate, placed side by side; but since the positive plate would only have a negative plate on one side of it, the other side being uncovered, it is preferable to associate two negative plates with one positive plate, so as to utilize the entire surface of the positive plate. Such a cell is shown in Fig. 119, where two negative plates, N, are placed one on each side of a single positive plate P, inside a jar $J\dot{}J\dot{}J$, filled to a convenient height with sulphuric acid and water. The positive plate is shown separately at A. It consists of a grid or frame of *antimonious lead;* i. e., lead alloyed with a small amount of antimony, so as to keep it from being acted upon by the acid liquid. In this frame are shown eight small circular buttons a, filling circular holes in the grid. These, when the cell is charged, consist of peroxide of lead, and constitute the active material of the cell. The negative grids have apertures filled with square buttons, which when in the charged condition are filled with porous

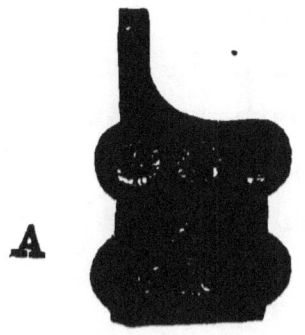

Fig. 119.—Small Storage Cell.

lead. In the cell shown, the plates are three inches square, and the weight of the entire cell, filled with solution, is four pounds. The capacity of this cell is about 6 ampere-hours when discharged at normal rates.

In a fully charged storage cell, the materials filling the apertures in the grids are dissimilar; namely, peroxide of lead and metallic lead. During discharge chemical actions occur, which result in reducing these substances to the same substance; namely, monoxide of lead. When now the charging current is sent through the exhausted cell, a dissimilarity is again produced, the monoxide being converted to peroxide of lead on the positive plate, and into porous lead on the negative plate. It is evident, therefore, that it is not electricity which is stored, but energy in the form

STORAGE BATTERIES. 357

of chemical energy, and in a form capable, on discharge, of being released under suitable conditions as electric energy.

Fig. 120, shows a larger size of the same type of storage cell. Here five positive plates are associated alternately with six negative plates, so that both external plates are negative; these plates are provided with rectangular apertures. The plates are bound together by an insulating frame F, F, F, and are individually separated or maintained at the proper distance apart by sheets of asbestos. In the cell shown, the plates are 10.5" square, and the entire cell, with solution, weighs 170 pounds. Its normal capacity is 500 ampere-hours, or, approximately, 3 ampere-hours per pound of total weight. All the positive plates are connected together to form a single positive plate with terminal P, and all the

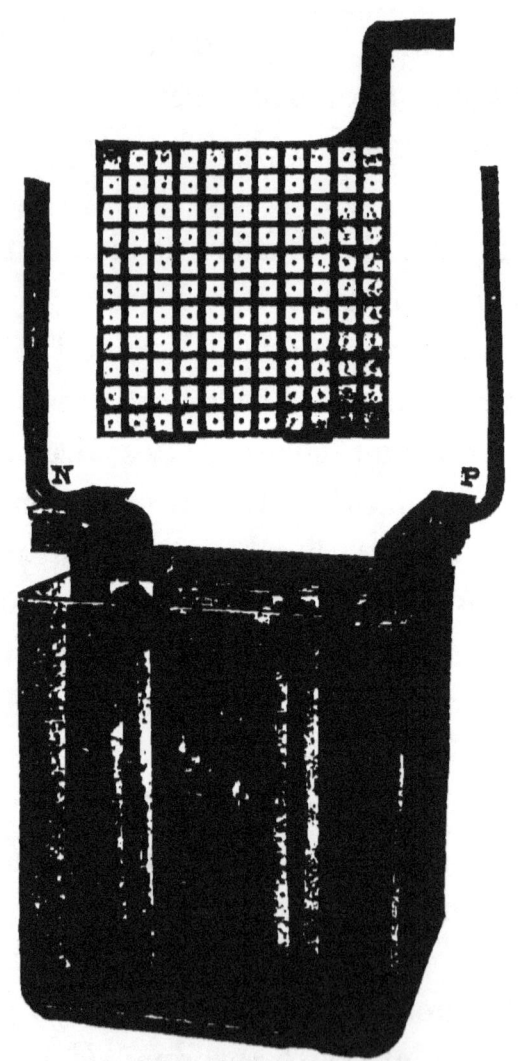

Fig. 120.—Storage Cell.

negatives are similarly connected to form a single negative terminal N.

When storage cells are employed in large central stations for the supply of powerful

FIG. 121.—LARGE STORAGE CELL.

currents, it is customary to associate large plates in single cells, rather than to employ a number of smaller cells in parallel. Fig.

121, represents a form of large cell for such an apparatus. Instead of employing a glass jar, the containing vessel is of wood, with an interior leaden lining. Here 21 negative plates are associated with 20 positive plates, each plate is 15.5" square, and the total weight of the cell, when filled with solution, is 800 pounds. The normal capacity of this cell is 5,000 ampere-hours, or about 6.4 ampere-hours per pound of total weight.

It is evident that the storage capacity increases with the weight of the cells; being over 6 ampere-hours per pound, with large cells, and with the smallest cells about 1 1/2 ampere-hours per pound of total weight. The E. M. F. produced by the average storage cell is about 2 volts. When fully charged, while discharging, it is about 2.2 volts, and falls during dis-

STORAGE BATTERIES.

charge to about 1.8. The total amount of energy which a storage cell supplies, expressed in watt-hours is, therefore, approximately, 2 multiplied by the capacity of the cell in ampere-hours. Thus the cell last mentioned has an *energy storage capacity* of 2 × 5,000 = 10,000 watt-hours, or 10 KW-hours, under normal circumstances.

Since incandescent electric lamps generally require a pressure of about 115 volts, and since a single storage cell has only 2 volts E. M. F., it is necessary to couple at least 57 cells in series, and, in order to allow for drop of pressure in the feeders, mains and house wires, the usual number required for this purpose is from 60 to 65 cells. For three-wire installations, twice this number is necessary, or about 120 cells. Such a three-wire storage battery

Fig. 122.—Central Station Storage Battery Installation.

installation is represented in Fig. 122. The cells are arranged in series, the positive terminal of one cell being connected to the negative terminal of the next. Each cell has 11 plates, 5 positive and 6 negative, and the normal capacity of the cell is 1,000 ampere-hours, giving a 10-hour discharge at the rate of 100 amperes.

A storage battery plant requires for its convenient operation a switchboard with its accessories, such, for example, as is shown in Fig. 123. $S, S,$ are two double-pole switches, one for controlling the charging, and the other the discharging circuit. The charging here being performed, as is usual in such cases, by a dynamo whose pressure is controlled by the *field rheostat* operated by a handle at F. $A, A,$ are ammeters in the charging and discharging circuits; $V,$ is a voltmeter which by means of the *pressure*

364   ELECTRIC INCANDESCENT LIGHTING.

FIG. 123.—STORAGE BATTERY SWITCHBOARD.

switch P, can be connected either to the dynamo terminals, or to the battery terminals, to show the pressure before the switch is thrown. T, is an *automatic cut-out* which disconnects the charging circuit of the dynamo, as soon as the C. E. M. F. of the cells exceeds the E. M. F. of the dynamo, and *O*, is an *overload switch* in the discharging circuit, arranged automatically to break the circuit of the battery should the discharge become excessive. *R*, is a switch which may either be used to throw in and out reserve cells, in order to maintain the discharging pressure constant, or to throw in and out *C. E. M. F. cells*, that are employed in place of the rheostat in the lamp circuit during charge.

The overload switch is shown in greater detail in Fig. 124. Here a coil of heavy wire *C,*—pivoted upon its axis, has its ends dipping in mercury cups. The coil is placed in the main circuit of discharge in such a manner that when the discharging current becomes excessive, the electromagnetic force developed by the coil rotates the helix and lifts the contact points P, P,

Fig. 124.—Overload Switch.

out of the mercury cups. H, is the handle for restoring the circuit when desired.

Fig. 125, shows a form of *overload switch* suited for heavier currents. The spring jaws, J, J, are in contact with the plates T, T, when the switch is closed, and the discharging circuit established. S, S, are

STORAGE BATTERIES. 367

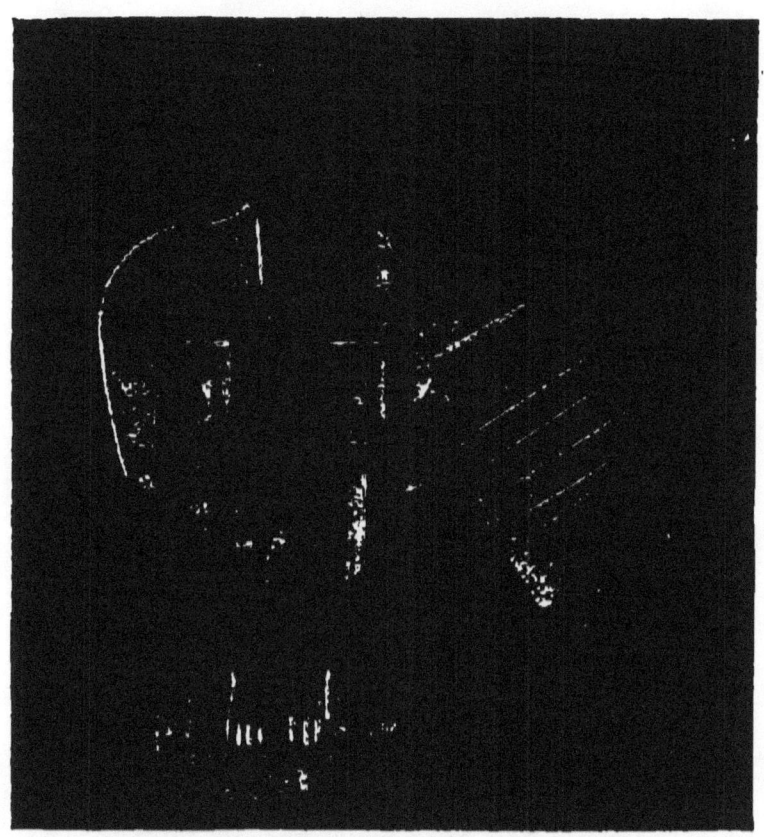

FIG. 125.—OVERLOAD SWITCH, CLOSED.

spiral springs which tend to disengage the plates and open the switch, but whose action is prevented by the armature *A*, which normally holds the switch plates in position. The coil C, of heavy conductor, is placed in the discharging circuit, and, as soon as the current strength exceeds a certain safe limit, it attracts the armature A, to the iron pole piece P, thus allowing the springs S, S to act, and the plates T, T, to be thrown out of the jaws J, J, as shown in Fig. 126.

In order to maintain a storage battery in proper working order, it is necessary to test the individual cells from time to time, for E. M. F. This is readily effected in practice, by means of any suitable voltmeter connected with electrodes for application to each separate cell. A convenient form of apparatus for this purpose is shown in Fig. 127. A simpler form of testing apparatus, not so accurate as the preceding, is shown in Fig. 128. It consists of a pair

Fig. 126.—Overload Switch, Released.

of handles with sharp metallic points, to make connection with the leaden electrodes of the cells, and connected by a loop of

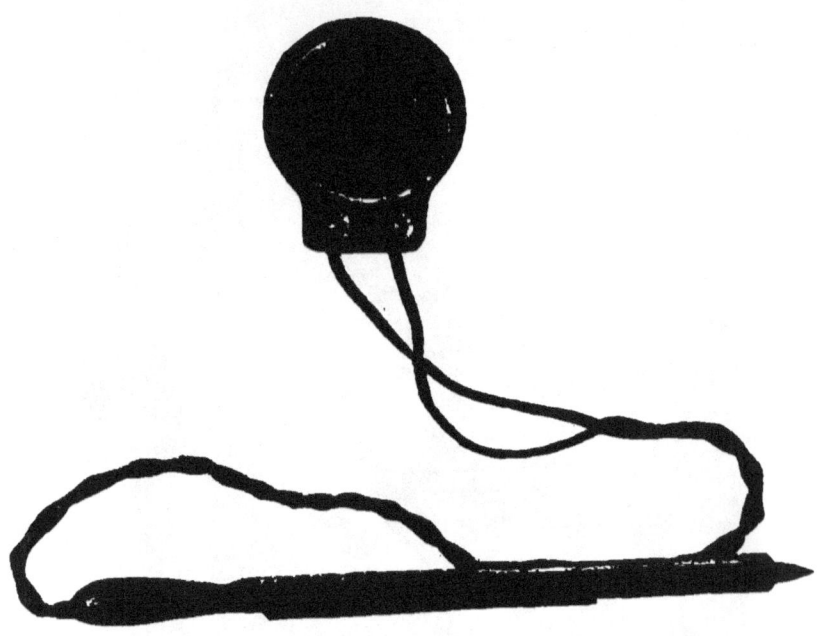

Fig. 127.—Storage Cell Voltmeter and Electrodes.

wire. In the end of one of the handles is a lamp socket in which a low volt lamp is inserted. If the cell tested is in good

Fig. 128.—Simple Form of Storage Cell Tester.

order its E. M. F. will be sufficient to bring the lamp up to candle-power.

The *efficiency* of a storage cell or battery is the ratio of the output to the intake. An ideally perfect battery would lose no energy, and would, therefore, have the same intake and output, representing an efficiency of unity, or one hundred per cent. In practice, the *ampere-hour efficiency* of a storage cell may be over ninety-five per cent. so far as regards electric quantity or ampere-hours, that is to say, if 100 ampere-hours be supplied to a storage battery during charge, it may yield more than 95 ampere-hours during discharge. On the other hand, since the pressure at the terminals of a cell during charge is over 2 volts, rising toward full charge to 2 1/2 volts, while the pressure during discharge is from 2.2 to 1.8 volts, the *energy efficiency*, as reckoned from the watt-hour output, is considerably less, and usually varies from seventy-five to eighty-five per cent. according to the conditions of the charge and discharge. A battery discharged at a rapid rate will not have so high an efficiency, either in quantity or in energy, as if slowly discharged. In relation, therefore, to coal consumed, the efficiency of a storage battery is about eighty per cent., but in relation to ampere-hours the efficiency may be over ninety-five per cent.

CHAPTER XVIII.

SERIES INCANDESCENT LIGHTING.

WE have have hitherto described multiple-connected lamps; or, in the case of the three-wire system, lamps in series-multiple. It sometimes happens, that a district to be lighted is scattered and wide spread, so that scattered lamps have to be supplied at considerable distances from the central station. As we have already shown, under these circumstances a very great amount of copper would require to be employed in their multiple-distribution. In order to avoid this, a series distribution is sometimes adopted. Here, as we have already described, a number of separate lamps are connected in series in one circuit. Since in such a system of distribution the extinguishment of a single lamp would open the entire circuit, a simple *automatic safety device* is required, which will establish a short circuit about the faulty lamp, in case it breaks, thus preserving the continuity of the circuit.

A system of series-distributed incandescent lamps may be operated from a constant high-potential dynamo. Thus, a dynamo of 1,000 volts E. M. F. may operate a circuit of 30 incandescent lamps, each having a pressure of 33 1/3 volts, including drop in connecting wires. Several such 1,000-volt circuits may be arranged in parallel, thus producing a multiple-series system of distribution, as shown in Fig. 129.

Such a series-connected system is some-

376 ELECTRIC INCANDESCENT LIGHTING.

times used for incandescent lamps in street lighting. It is not suitable for house lighting, since it is essential that the number of lamps in each series circuit should

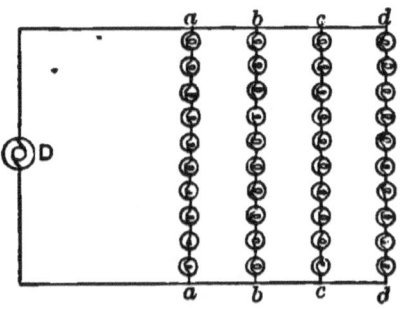

129.—MULTIPLE-SERIES SYSTEM.

be as nearly constant as possible, and this precludes the possibility of cutting lamps out of a circuit when no longer required.

Various forms of automatic circuit-protecting devices have been produced. One of the simplest of these, called the *film cut-out,* consists of a thin strip of paper placed between two spring contacts connected with two terminals of each lamp. While the lamp remains lighted, the electric pressure across the thickness of the strip of paper is only a few volts; *i. e.*, the pressure at the lamp terminals, but if the filament should break, the pressure at the terminals is the full pressure of the circuit which may be 1,000 volts. This pressure is capable of piercing the thin paper sheet, thus establishing an arc, which instantly welds the two metal strips together, thus short-circuiting the lamp and re-establishing the circuit.

As already stated, series incandescent lamps are generally employed for out-door lighting and, therefore, have to be protected from the weather. The comparatively high electric pressure employed on these circuits, not being safe to handle under all conditions, requires certain precautions in insulation when introduced into buildings. Fig. 130, shows a form of lamp and fixtures suitable for out-door series-incandescent lighting. The lamp L,

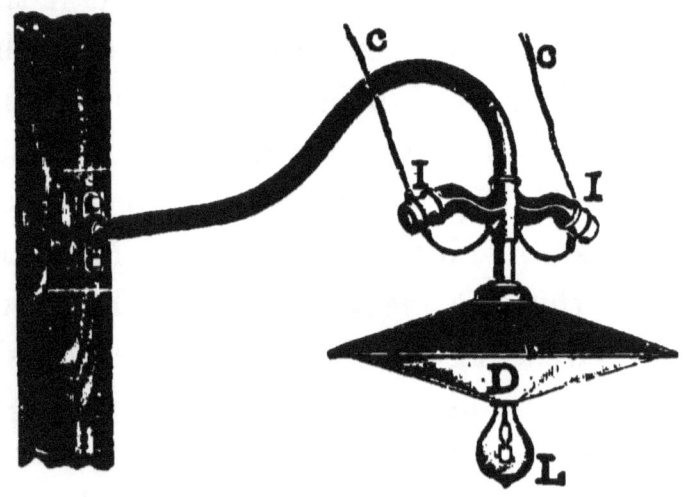

FIG. 130.—STREET FIXTURE, WITH SERIES INCANDESCENT LAMP.

is placed in its socket beneath the porcelain deflector D, which not only serves to scatter and reflect the light, but also, in conjunction with the cover C, to protect the

SERIES INCANDESCENT LIGHTING. 379

lamp socket from rain. The circuit wires c, c, are brought to the socket of the lamp after being secured to the insulators I, I. Fig. 131, shows the same fixture with the

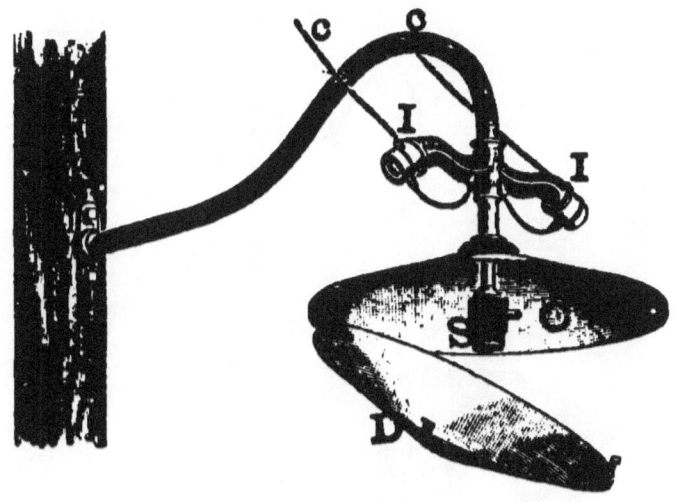

Fig. 131.—Street Fixture, with Lamp Removed and Socket Exposed.

deflector D, removed, showing the socket S, placed in the interior. Fig. 132 shows a form of lamp post suitable for street lighting with such lamps. Another form

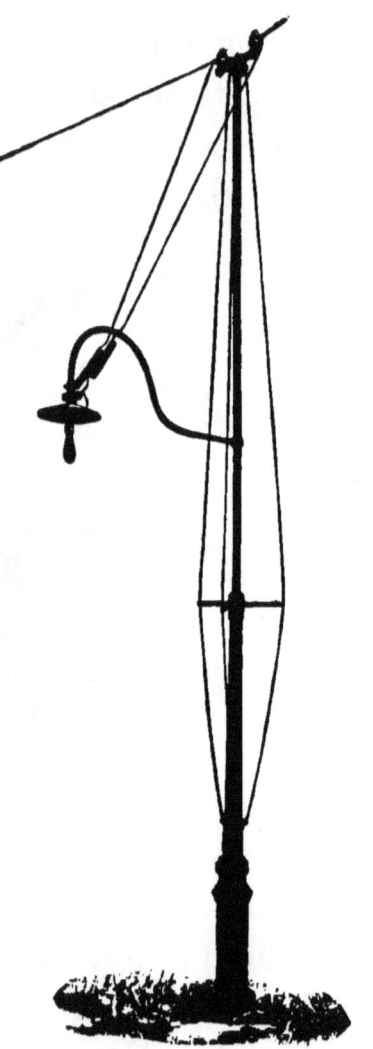

Fig. 132.—Lamp Post.

SERIES INCANDESCENT LIGHTING.

of fixtures is shown in Fig. 133; here the incandescent lamp is provided with an external globe.

Where a series circuit has already been installed for operating arc lamps, it is

FIG. 133.—STREET LAMP FIXTURE.

sometimes desirable to insert incandescent lamps in the same circuit. This is done by employing series incandescent lamps of special manufacture. Since the current strength, in a series-arc circuit, is usually about 10 amperes, and is the same in all parts of the circuit, it is necessary that the incandescent lamps be made for this current strength. A fifty-watt 16-candle power lamp, taking 10 amperes, must be a 5-volt lamp, since 5 volts × 10 amperes = 50 watts. Consequently, the resistance of the lamp must be only 1/2 ohm, since 10 amperes × 1/2 ohm = 5 volts, and the filament must be comparatively short and thick in order to possess this relatively low resistance. It is evident that lamps of different candle power can be obtained from the use of such series circuits, by suitably adjusting the resistance of their filaments. The connections for such a series arc and incandescent circuit are shown in Fig. 134.

Since series-arc circuits frequently employ dangerously high pressures, it becomes unsafe to come in contact with the conductors of such circuits when standing on wet ground; for, such circuits com-

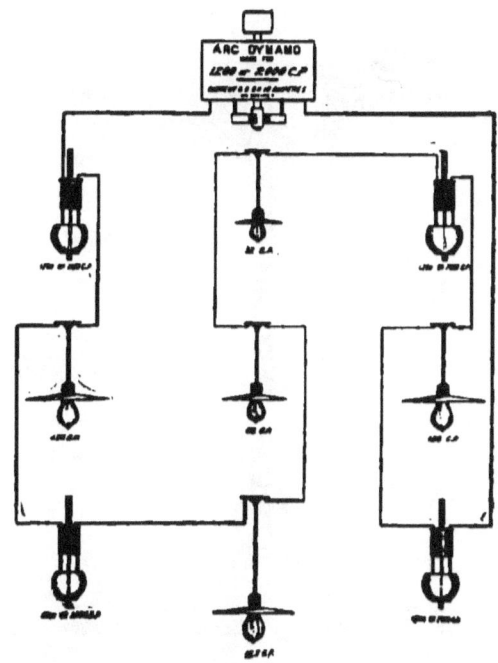

FIG. 134.—SERIES ARC AND INCANDESCENT CIRCUIT.

monly have marked leakage and thus a dangerously high current might be sent through the body. Consequently, it is

384 ELECTRIC INCANDESCENT LIGHTING.

necessary carefully to insulate the keys and connections of incandescent lamps operated on arc circuits.

Fig. 135.—Lamp for Series-Arc Circuit, with Cut-out Switch.

Fig. 135, shows a form of lamp suitable for series-arc circuits. Like all series-connected lamps, it contains an automatic cut-out, and of the film type already described. It has no key, but is provided with a ceiling switch whereby it may be short-circuited and thus extinguished.

CHAPTER XIX.

ALTERNATING-CURRENT CIRCUIT INCANDESCENT LIGHTING.

SINCE incandescent lamps are frequently operated on alternating-current circuits, it may be well briefly to discuss some of the characteristics of these currents and instances of their commercial application.

In a continuous current the direction of electric flow does not change. Its strength may vary periodically, in which case the current is said to *pulsate;* or, it may be unvarying, in which case the current is said to be *steady*. An alternating current, on the contrary, changes direction many

ALTERNATING-CURRENT CIRCUIT.

times in a second, there being first a wave or flow of current through the circuit in one direction, and then a wave or flow of current in the opposite direction, and so on. Each of these waves or flows is called an *alternation*, and a complete to-and-fro motion, or double wave, is called a *cycle*. The *frequency* of alternation is the number of alternations or of cycles executed in a second; thus ordinary commercial alternating circuits have a frequency of from 25 cycles, or 50 alternations per second, to 140 cycles, or 280 alternations per second. The number of cycles, or complete periods, is sometimes symbolized by the sign ∼, so that a frequency of 140 complete periods, or cycles, per second would be written 140 ∼.

Between each pair of successive waves, at the time when the current is changing direction, there is no current flowing in the circuit. Consequently, it might be supposed, when an alternating current, of say 120 ~, is sent through a lamp, since there would be 240 moments in each second when no current is flowing, that the light furnished by the lamp would pulsate, being extinguished and relighted 240 times in a second. In point of fact, this tendency to pulsate does exist; but when the frequency is sufficiently high, before the filament loses enough of its heat to cease glowing, it receives a fresh accession of heat from the succeeding wave of current. Moreover, the retina of the eye tends to retain its luminous impressions for a sufficiently great fraction of a second to aid in the apparent uniformity of the light. The result is, when the frequency is above say 30 ~, or 60 alternations per second, *i. e.*, 3,600 alternations per minute, that the light of an incandescent lamp is practically steady. As the frequency is reduced, the *higher economy lamps;* i. e., those which have a greater efficiency, or a greater number of candles-per-watt, are the first to suffer in apparent steadiness. First, because they are more brilliant for the same candle-power, or are operated at a high temperature, and the eye is much more sensitive to changes of brilliancy than to changes of candle-power, or, in other words, to changes in candle-power per square inch of bright surface, than to total candle-power; and second, because such lamps have usually thinner filaments, and hence chill more rapidly.

The frequency employed in alternating-current circuits varies between 25 ∼ and 140 ∼ per second.

If an incandescent lamp be supplied by a continuous pressure, of say 110 volts, and gives a candle-power of 16 candles at this pressure, then if it be removed and connected with an alternating-current pressure of 110 volts, it will shine with equal brightness and give 16 candles as before, no matter what the frequency, provided only that the frequency be sufficiently great to keep the lamp from flickering. Similarly, the current strength, which the lamp will take, on an alternating current circuit, is the same as that which it takes on a continuous-current circuit. This is for the reason that although the strength of an alternating current is constantly fluctuating, between the apex of the successive waves and zero at the moments of reversal, yet the current is defined or measured by its heating effect, so that if half an ampere of continuous current brings a given lamp to candle-power, then the alternating-current strength which also brings the lamp to candle-power, will be half an ampere, no matter what the outline of the current wave may be. Thus, it would be possible for each wave to have a current strength of 2 amperes at the apex, and yet to produce only the average heat effect of a half an ampere. Such a current would have an *effective current strength* of 1/2 ampere. In general, the maximum current strength is about forty per cent. in excess of the effective current strength, so that when an alternating-current ammeter shows that a current of one effective ampere is passing in a circuit, the apex of the successive current waves will, probably, attain a strength of $1\frac{4}{10}$ amperes. In the same way, if the pressure in an alternating-current circuit, as shown by lamps or by instruments, be 100 volts, the instantaneous maximum value in the successive cycles will, probably, be about 140 volts.

It is evident, therefore, that it would be possible to employ alternating-current generators instead of continuous-current generators, in a central station, for the distribution of electric light. The drop of pressure, however, in the supply feeders and mains, would, in such a case, for reasons that it is not necessary here to consider, be greater than with the same strength of continuous current, and it is generally considered, that it is disadvantageous to supply low-tension systems of incandescent lighting from alternating-current generators. When, however, the lighting has to be distributed at great distances, and, therefore, at a high electric pressure, in order to avoid expense in conductors, the alternating-current system possesses decided advantages, since it readily lends itself to transformation of pressure; that is to say, it is easy to take an alternating-current generator of say 2,000 volts E. M. F., and to transmit this pressure to considerable distances over comparatively small conductors, and then to reduce the pressure locally, within the precincts of buildings, to say 100 volts, by means of an apparatus, called an *alternating-current transformer*.

A form of alternating-current transformer is shown in Fig. 136. Such an apparatus consists essentially of two coils of insulated wire, wound around a common laminated iron core. These coils are connected respectively with a high-pressure and a low-pressure circuit. In this case, the high-pressure circuit is the source of energy, or the *primary circuit*, and the low-pressure circuit is the circuit of delivery, or the *secondary circuit*. The effect of sending an alternating current through the primary coil is to induce alternations of

FIG. 136.—ALTERNATING-CURRENT TRANSFORMER.
CAPACITY 1 KW, OR 20 50-WATT LAMPS.

the same frequency, but different E. M. F., in the secondary coil. If the secondary coil contains more turns than the primary, then the secondary E. M. F. is the greater. If, on the contrary, as in this case, the secondary turns are fewer, then the secondary E. M. F. will be lower. Transformers are, therefore, of two kinds; namely, *step-up transformers*, or those which raise the pressure, and *step-down transformers*, or those which lower it.

Like all machines for effecting the transformation of energy, alternating-current transformers waste some portion of what they receive. In large transformers this loss at full load is relatively very small, only about two per cent., so that the efficiency of a large transformer, at full load, is approximately ninety-eight per cent. On the other hand, the relative loss at very light load is necessarily much greater. During the day time, when comparatively few lamps are lighted over the distribution system, the power supplied to magnetize the transformers may be considerably greater than the power usefully expended. This is the only objection to the action of alternating-current transformers, and is outweighed when the distance to which the current has to be carried is sufficiently great, since, otherwise, a large amount of copper would have to be employed to transmit the necessary energy in any other way. A continuous current cannot be transformed without the aid of rotating mechanism.

In Fig. 136, P, P, are the primary wires leading to the high-pressure mains, which are usually supported on poles overhead. S, S, are the secondary wires leading to the interior of the building and acting as service wires. The apparatus represented in the figure has a capacity of 1 KW, and, therefore, is capable of supplying about 20 fifty-watt incandescent lamps. If the secondary pressure be 100 volts, the current strength at full load would be 10 amperes approximately, since 100 volts × 10 amperes = 1,000 watts. In the primary circuit the current strength will be about 1 ampere at full load, if the primary pressure be 1,000 volts, since 1,000 volts × 1 ampere = 1,000 watts. In reality, owing to some loss of power in the transformer, say 50 watts, the current strength would be somewhat in excess of 1 ampere. Moreover, in alternating-current circuits, the current strength is in excess of the number of amperes which, multiplied by the pressure in volts, gives the actual activity in watts.

Transformers of small size, weight and capacity, are more expensive per KW, or per lamp, both to purchase and to operate than large transformers, since they require relatively more labor and material and have a lower efficiency. It is customary, therefore, when possible, to supply several sets of secondary conductors, say in several adjacent buildings from a single large transformer, rather than have a transformer for each house. For the same reason, where single lamps have to be supplied by alternating currents at considerable distances apart, as, for example, in street lighting, instead of employing a small $\frac{1}{20}$th KW transformer for each lamp, the method is adopted of connecting a number of lamps in series, as in an arc circuit, and shunting each lamp by a coil called a *reactive coil*. This reactive coil allows almost the entire current strength in the circuit to pass through the lamp and takes only a small portion through its own circuit. If, however, the lamp filament breaks, thus interrupting the circuit of that lamp, the reactive coil permits the full current strength to pass through it with only a comparatively small drop in pressure; namely, the pressure

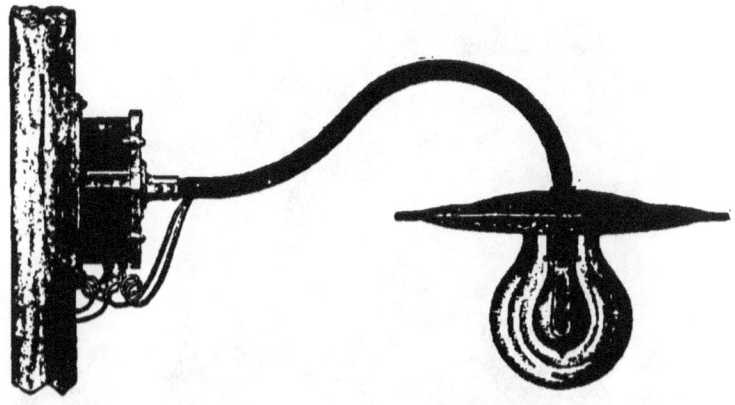

FIG. 137.—SERIES-CONNECTED STREET LAMP FOR ALTERNATING-CURRENT CIRCUITS.

equal to that of the lamp which has become extinguished. This drop is due to a C. E. M. F. established by the current in the circuit in passing through the reactive coil. Such a series lamp shunted by its reactive coil is represented in Fig. 137.

The alternating currents required for alternating-current incandescent lighting,

FIG. 138.—ALTERNATING-CURRENT GENERATOR.

are obtained from a form of dynamo known as an *alternator*. This dynamo does not differ in principle from an ordinary continuous-current generator, except that it does not commute the current in its line circuit. Fig. 138, shows a form of alternator having fourteen poles. Since the fourteen field magnet coils require to be supplied by continuous currents, the machine is accompanied by a small continuous-current generator, called an *exciter*.

Incandescent lamps, intended for alternating-current circuits, possess no peculiarity, that is to say, a lamp may be used indifferently on a continuous or on an alternating-current circuit, when the working pressures in each case, and the sockets, are the same.

CHAPTER XX.

MISCELLANEOUS APPLICATIONS OF INCANDESCENT LAMPS.

BESIDES the various uses for the incandescent lamp, which we have described, there are many others which want of space will prevent our discussing, except very briefly.

The fact that the incandescent lamp is capable, within reasonable limits, of being made of almost any size, and the additional fact that the glowing filament is entirely protected by a surrounding glass chamber, permits the incandescent lamp to be employed for purposes in which other artificial illuminants would be impossible.

We may mention, as one of such purposes, the employment of small suitably shaped incandescent electric lamps for exploring the cavities of the body. Incandescent lamps are thus employed by physicians and surgeons. For this purpose a small lamp, shaped so as to permit of its ready introduction into the cavity to be examined, is mounted at the extremity of a suitable support. In some cases the exploring lamp is sometimes placed in a sheath or tube, through the interior of which the illumined area is directly observed.

An entirely distinct method, however, from the preceding is sometimes adopted; namely, the method of *trans-illumination.* Here an attempt is made to illumine the interior cavity so as to permit it to be visible through the body as a translucent screen.

404   ELECTRIC INCANDESCENT LIGHTING.

The amount of light required to attain this result, is so great as to necessitate the liberation of considerable heat activity, and means have usually to be provided for keeping the lamp cool. This is done by a

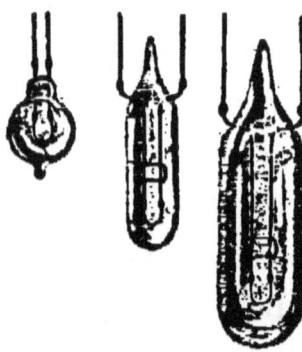

Fig. 139.—Miniature Incandescent Lamps.

jacket, in the exterior glass globe, through which cold water is kept circulating.

Examples of three small lamps employed for surgical and dental purposes are seen in Fig. 139. Such lamps are generally called *battery lamps*, because they are cap-

MISCELLANEOUS APPLICATIONS. 405

able of being operated from a primary or secondary battery. The usual voltage is from three to six volts. The efficiency of such lamps, in candle-power per watt, is

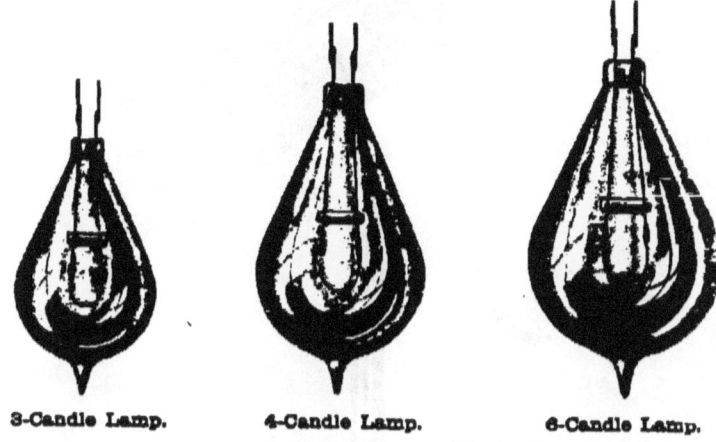

3-Candle Lamp.   4-Candle Lamp.   6-Candle Lamp.

FIG. 140.—MINIATURE INCANDESCENT LAMPS.

much less than that of larger lamps, owing principally to the rapid conduction of heat from the short filament through the conducting wires. Other forms of battery lamps, of candle-power as marked, are represented in Fig. 140.

Sometimes in medical or surgical examinations, a small incandescent lamp is mounted before a concave reflector, strapped

FIG. 141.—INCANDESCENT LAMPS, WITH REFLECTOR AND CONDENSING LENS.

to the head of the observer. In other cases the lamp and reflector are mounted as shown in Fig. 141, within a tube mounted on a suitable stand and provided with a condensing lens.

Another use for the incandescent electric lamp which is due to the fact that the glowing filament is hermetically sealed, is the use of *safety incandescent lamps* in mines, where the presence of fire-damp is feared. This form of safety lamp is intended to replace the Davy safety lamp. Such lamps are supplied by either a primary or secondary battery placed in the lamp case. A form of such lamp is shown in Fig. 142.

Battery or miniature incandescent lamps are also often employed for use with the microscope in the illumination of the object, since they are capable of being placed conveniently for observation. Various designs of supports and reflectors are

408   ELECTRIC INCANDESCENT LIGHTING.

employed in connection with lamps for this purpose.

The incandescent lamp is used exten-

Fig. 142.—Safety Lamp for Mines.

sively on the modern steamship not only for purposes of general illumination of the vessel, but also for the lighting of the and *signal* ~~lights. In~~ the case of such lights, it is a matter of considerable importance that the continuity of service of the lamps be ensured, since the rupture of the filament might jeopardize the vessel. In order to lessen the liability of disabling the side-light, it is common to employ a *double-filament lamp*, sometimes called a *twin-filament lamp*, so arranged that if one filament should fail, the other may continue burning. At other times two separate lamps are employed for the same purpose. A double-filament lamp is shown in Fig. 143.

Incandescent lamps on board ship are operated at pressures between 100 and 120 volts, almost invariably on the two-wire system. In warships, the pressure is usually 80 volts, in order to facilitate the use

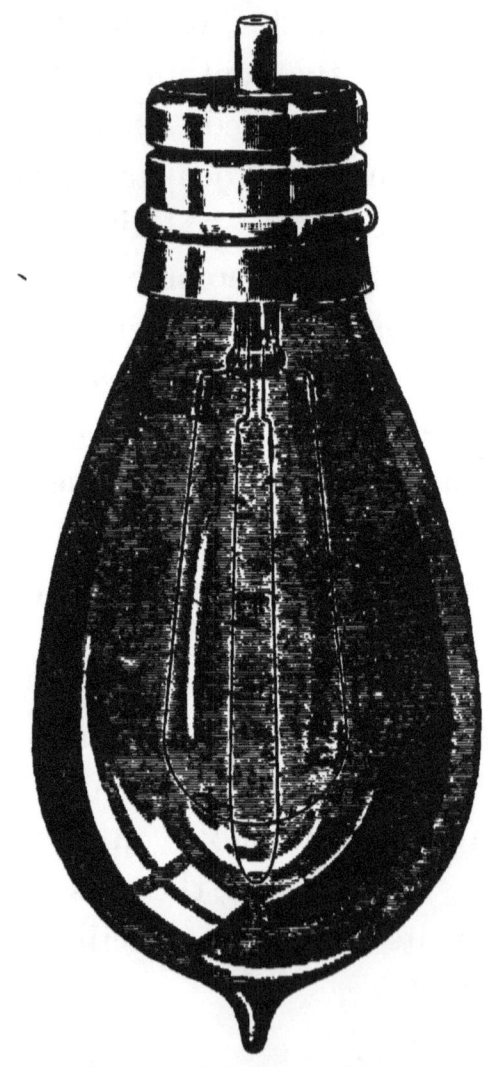

Fig. 143.—Double Filament Lamp for Ship's Side-light.

of arc-light projectors without the addition of much resistance in the projector circuits.

In the early history of ship lighting, the method was adopted of employing the ship's iron sheathing as a common return; that is, one pole of all the lamps and also one pole of the dynamo being connected to the sheathing of the vessel, the other pole being connected to insulated conductors. This method was, however, found objectionable in practice and has been completely given up. Especial care has to be given to the insulation of wires and fixtures on board of ship. The connection boxes, switches, etc., are often arranged so as to be rendered completely water-tight. In Fig. 144, H, is the switch handle, and R, the receptacle for the insertion of local connecting wires, leading to the lamp or other device to be operated. The cover

412   ELECTRIC INCANDESCENT LIGHTING.

C, is intended to effect a water-tight seal when the receptacle is removed. In Fig.

Fig. 144.—Water-Tight Marine Switch and Receptacle.

145, the switch handle *H*, is recessed into the water-tight box in such a manner that the cover *C*, can enclose it in a water-tight seal. The wires enter the box through the water-tight openings at O O.

Miniature incandescent lamps are occasionally employed as a variety of *electric*

Fig. 145.—Water-Tight Marine Switch.

jewelry, as in a scarf pin or ornament for the hair, such lamps being always operated from a small battery carried on the person. Similar lamps are also employed on the stage, for stage effects, being worn by the actors.

Incandescent lamps are colored either by employing globes of tinted glass, or by dipping the clear lamp globes into a solution of suitable dye. In the former case the coloring is permanent, in the latter case, it is temporary.

Incandescent lamps, either plain or colored, are extensively employed in *illuminated electric signs*, where the lamps are grouped in the shapes of letters. Sometimes, in order to attract attention to the sign, automatic switch devices are employed, which at intervals extinguish some or all of the lamps; or, blocks of lamps are arranged so as to be automatically cut out of circuit and cause letters to successively appear and disappear and words to be thus spelled out.

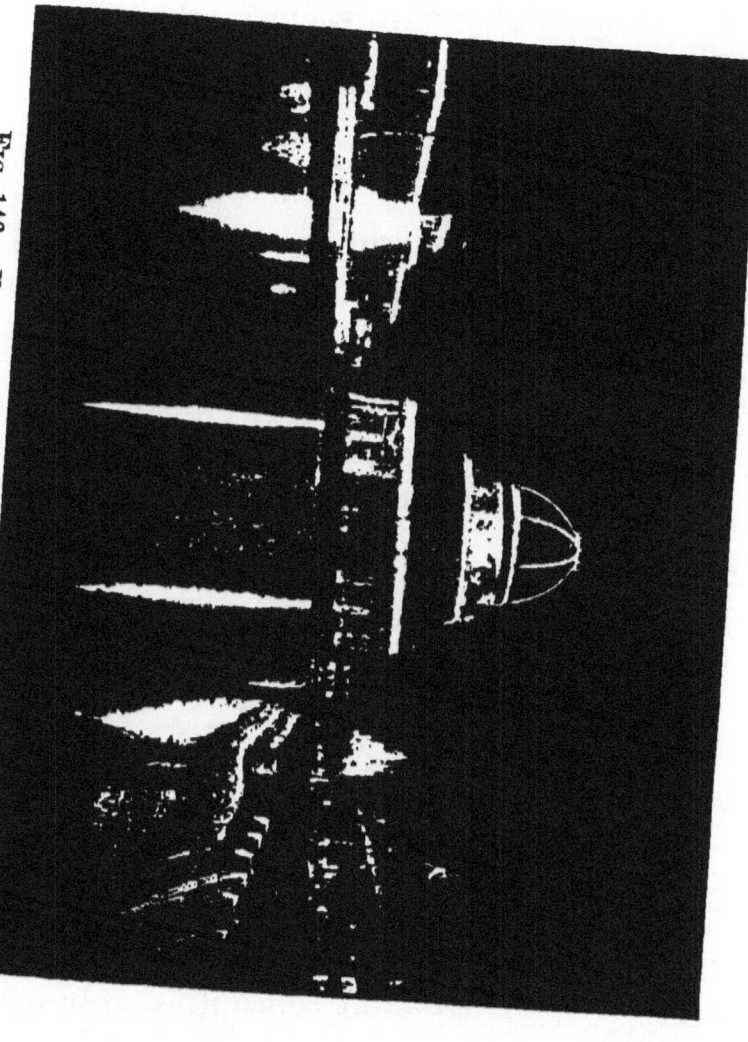

FIG. 146.—ELECTRIC LIGHTING AT THE WORLD'S FAIR.

The incandescent electric light lends itself very readily to decorative effect. This arises not only from the readiness with which the light is subdivided and distributed, but also from the fact that the glowing filament being protected by the surrounding glass chamber, permits the light to be partially buried in walls or ceilings, in a manner which would be impossible with any other form of artificial illuminant.

No description of decorative effect by incandescent lamps would be complete without some allusion to the magnificent spectacle that was afforded by the incandescent lighting of the Court of Honor, at the World's Columbian Exhibition, at Chicago, in 1893. A faint conception only of the beauty of this scene may be gathered from the accompanying illustration in Fig.

146. The building in the background is the Administration Building, whose exterior is lighted by incandescent lamps. On the right hand side is the façade of Electricity Building, and on the left hand is the façade of Machinery Hall. The entire bank of the waterway was illumined by incandescent lamps.

INDEX.

A

Actinic Effect of Light, 208.
Activity, 62.
——— and Candle-Power, Relation Between, 183.
———, Surface, of Incandescent Filament, 165.
———, Surface, of Positive Crater in Arc. 166.
———, Unit of, 63.
Adapters for Lamps, 133, 134.
Adjustable Lamp Pendant, 240.
Age Coating of Lamp, 178.
Air Pump, Mechanical, 125.
Alternating-Current Circuit for Incandescent Lighting, 386 to 401.
——— Current Generator, 400.
——— Current Transformer, 393.
Alternation, 386.
———, Frequency of, 387.
Alternator, 400.
Ammeters, 306.

INDEX.

Ampere, 56.
——— Hours, 335.
——— Efficiency of Storage Cell, 372.
Amyloid, Use of, for Filaments, 87.
Analysis of Light of Glowing Filament, 74.
——— of Sunlight, 74.
Antimonious Lead, Use of, in Storage Cells, 354.
Arc, Voltaic, 18.
Armature for Central-Station Generator, Winding of, 322, 323.
——— of Central-Station Generator, 314.
Artificial Illuminants, Requisites for, 6.
——— Illumination, 1 to 17.
Automatic Cut-Out for Storage-Battery Switchboard, 365.
——— Safety Device for Incandescent Lamp, 375.
——— Switch, 273.

B

Bamboo Filament, 107.
——— Filament, Preparation of, 86, 87.
Bases, Lamp, 131, 132.
Batteries, Storage, 345 to 373.
Battery Incandescent Lamps, 407.
——— of Boilers, 311.
———, Voltaic, 47.

Begohm, 247.
Belt-Driven Generator, 327.
Belting, 327.
Bipolar Generator for Isolated Plant, 330.
Blackening of Lamp Chamber, 178.
Block, Branch, 274, 275.
Blowing of Fuse, 276.
Boilers, Battery of, 311.
Bougie-Decimale, 200.
Boulyguine, 33.
Boulyguine's Incandescent Lamp, 32, 33.
Box, Carbonizing, 93.
Boxes, Distribution, 280, 282.
———, Connecting, 263, 264.
———, Coupling, 295, 296.
———, Field Regulating, 306.
———, Junction, 300 to 303.
Bracket, Lamp, 243.
——— Lamp, Movable Arm for, 246.
Branch, 274, 275.
——— Coupling Boxes, 297.
——— Cut-Out, 274.
Branches, 251.
Brass-Covered Conduit, 262.
Brilliancy, 181.
——— of Incandescent Filament, 168.
British Candle, 199.

Brushes for Generator, 316.
Bus Bars, Definition of, 223.

C

Candle Power and Activity, Relation Between, 183.
——— Power, Effect of Varying Pressure on, 188 to 191.
——— Power of Luminous Source, 199.
——— Power, Mean Spherical, 207.
Capacity, Energy of, Storage of Cell, 361.
———, Storage, 349.
Carbon Filament, Life of, 167.
———, Suitability of, for Incandescent Filament, 83, 84.
Carbonization, General Processes for, 84, 85.
———, Methods Employed for, 91 to 98.
Carbonizing Box, 93.
——— Frame, 94, 95.
Cell, Charged, 352.
———, Counter-Electromotive Force of, 365.
———, Secondary, 352.
Celluloid Filaments, 90.
Central-Station Generator, Double Brushes for, 316.
——— Station Generators, 313.

INDEX. 423

Central Station, Load Diagram of, 347, 348.
—— Station, Six-Pole Generator for, 317 to 320.
—— Station Smashing Point of Lamp, 193.
—— Station, Storage Cell for, 359.
—— Station Switchboard, 305 to 308.
Central Stations, 304 to 333.
Charged Cell, 352.
Charging Current, 352.
Chlorine and Bromine, Residual Atmospheres of, in Lamp Chambers, 130.
Circuit, Open, 46.
——, Primary, of Transformer, 393.
——, Secondary, of Transformer, 394.
——, Series Arc and Incandescent, 383.
Circular-Mil, Definition of, 54.
Circular-Mil-Foot, 54.
Cleat Wiring, 254.
Cleats, 255.
——, Wooden, 255.
Coil, Reactive, 398.
Cold Light, 10, 78, 79.
Color, Cause of, 71, 72.
—— Values, Day-Light, 72.
Commutator of Central-Station Generator, 314.
——, Sparking at, 317.
Concave Panel Shade and Reflector, 153.
Concealed Work, 260.

Conduction of Heat, 75.
Conductor, Neutral, of Three-Wire System, 222.
———, Twisted-Double, 248.
Conductors, Double-Flexible, 249.
———, Drop in, 252.
———, Parallel, 248.
———, Silk-Covered, 248.
———, Solid, 247.
———, Supply, 250.
———, Twin, 248.
———, Stranded, 247.
Conduit, Brass-Covered, 262.
——— of Creosoted Wood, 288, 289.
Conduits, 287.
———, Interior, 261, 262.
Connecting Boxes, 263, 264.
Connection Boxes for Interior Conduits, 264.
——— in Series, 47.
Consumers' Smashing Point of Lamp, 194.
Continuous Electric Current, 304.
Convection of Heat, 77.
Cord, Flexible Lamp, 240.
Corrugated Lamp Reflector and Shade, 155.
Cotton Thread, Use of, for Filament, 87.
Coulomb, 56.
Coulomb-per-Second, 56.
Counter-Electromotive Force Cells, 365.

Coupling Boxes, 295.
———— Boxes, Branch, 297.
Crater, Positive, of Arc, Surface Activity in, 166.
Creosoted Wood Conduit, 288, 289.
Current, Continuous Electric, 304.
————, Electric, Pulsating, 386.
————, Steady Electric, 386.
———— Strength, Effective, 391.
Cut-Out, Branch, 274.
———— Film for Incandescent Electric Lamp, 376, 377.
———— Fixture, 278, 279.
———— Mains, 277.
———— Switch for Series Circuit, 384.
Cycle, 387.

D

Day-Light Color Values, 72.
De Changy, 25.
———— Incandescent Lamp, 26, 27.
De Moleyn, 24.
Diagram of Central-Station Load, 347, 348.
Direct-Driven Generator, 328, 329.
Distributing Point, 280.
Distribution Boxes, 280, 283.

Distribution, Centers of, 253.
———, Series-Multiple Lamp, 221.
———, Three-Wire System of Lamps, 221 to 224.
Double Brushes for Central-Station Generator, 316.
——— Filament Lamps, 409.
——— Flexible Conductors, 249.
——— Pole Switches, 267.
——— Pole Switches, Forms of, 269, 270, 271.
Drop in Conductors, 252.
Dummy Moulding, 260.
Dynamo-Electric Generator, 48.
Dynamos or Generators, 311.

E

E. M. F., 44.
Early History of Incandescent Lighting, 18 to 42.
Early Horse-Shoe Lamp, 104.
Early Illuminants, 1 to 5.
——— Incandescent Lamps, 19, 20.
Effect of Temperature on Resistivity of Insulators, 55.
Effective Current Strength, 391.
Efficiency, Ampere-Hour, of Storage Cell, 372.

Efficiency of Incandescent Lamp, 170 to 233.
―――― of Lamp, Effect of, on Duration of Life, 184, 185.
―――― of Storage Cell, 372.
―――― of Transformers, 395.
Electric Jewellery, 413.
―――― Lighting, Life Risks of, 14, 15.
―――― Pressure, 45.
―――― Quantity, Unit of, 56.
―――― Resistance, 49.
Electrolier, 243.
Electrolytic Meter, 335 to 339.
―――― Seal, 108.
Electromotive Force, 43.
Elementary Electrical Principles, 43 to 64.
Elements or Plates of Storage Cell, 351.
Emissivity of Incandescing Filament, 80, 81.
―――― of Lamp Filament, Effect of Surface on, 180.
Energy Efficiency of Storage Cell, 372, 373.
―――― Storage Capacity of Secondary Cell, 361.
―――― Storage of Cell, 361.
Equalizer Switch, 235, 236.
Ether, Luminiferous, 67.
――――, Universal, 67.
Evaporation of Incandescent Filament, 176, 177.
Exciter, 401.

F

Factor, Load, 349.
Farmer, 36.
Farmer's Incandescent Lamp, 35, 36.
Feeder Distribution, 229.
—— Distribution, Three-Wire System of, 231, 232.
—— Equalizer Resistance, 234.
—— Load, Methods of Overcoming Inequalities of, 233, 236.
—— Regulators, 233.
—— System, 253.
—— Tubes, 298, 299.
Feeders for Lamp Distribution, 228.
Feeding Point, 298.
Field Magnet Coils of Generator, 314.
—— Regulating Boxes, 306.
Filament, Bamboo, Preparation of, 86, 87.
———, Effect of Flashing on Emissivity of, 116, 117.
———, Effect of Surface on Emissivity of, 180.
———, Incandescing, Surface Activity of, 80, 81.
———, Mounting of, 99, 100.
———, Sealing-in of, 118.
———, Shadows, 179.
———, Spotted, 112.

Filaments, Amyloids, 85.
———, Celluloid, 90.
———, Flashing Process for, 113, 114, 115.
———, Methods Employed for Carbonization of, 91 to 98.
———, Squirted, 88, 89, 97.
———, Stopper-Mounted, 122, 123.
———, Use of Cotton Thread for, 87.
Fire Fly, Radiation of, 78.
Fittings, Lamp, 135 to 162.
Five- and Four-Wire Systems of Lamp Distribution, 225.
Fixture Cut-Outs, 278, 279.
——— Molding, 259.
——— for Street Incandescent Lamp, 381, 382.
Fixtures and Wiring for Houses, 237 to 283.
Flashing Process for Filaments, 113, 114, 115.
Flexible Lamp Cord, 240.
——— Lamp Pendant, 239, 240.
Flush Switches, 272.
Foot-Pound, 59.
Foot-Pound-per-Second, 63.
Four- and Five-Wire Systems of Lamp Distribution, 225.
Frame, Carbonizing, 94, 95.
French Standard Candle, 200.
Frequency, Effect of, on Steadiness of Light, 388.

Frequency of Alternation, 387.
——, Luminous, 68.
Full-Wire Guard for Incandescent Lamp, 158.
Fuse, Blowing of, 276.
——, Cut-Out, 273.
——, Safety, 274.

G

Geissler Type of Mercury Pump, 127.
Generator, Alternating-Current, 400
——, Dynamo-Electric, 48.
——, Field Magnet Coils of, 314.
—— Unit, 308.
Generators, Multipolar, 316.
Generators or Dynamos, 313.
Glass Lamp Shades, 156.
Glow Worm and Fire Fly, Radiation of, 78.
Glowing Filament, Analysis of Light Produced by, 74.
Grids of Storage Cell, 351.

H

Half-Shade for Incandescent Lamp, 151.
Half Wire-Guard for Incandescent Lamp, 157.
Heat, Conduction of, 75.
——, Convection of, 77.

Heat, Molecular Transfer of, 76, 77.
———, Radiation of, 75.
High-Economy Lamps, 389.
Horizontal Intensity, Maximum, 207.
Horse-Power, Definition of, 63.
Hours, Ampere, 335.
———, Watt, 335.
House Fixtures and Wiring, 237 to 283.

I

Illuminants, Early, 1 to 5.
Illuminated Electric Signs, 414.
Illumination, Actual Values of, 206.
———, Artificial, 1 to 17.
———, Law of, 203, 204, 205.
———, Significance of, 198.
———, Unit of, 202.
Incandescent Lamp, 163.
——— Lamp, Automatic Safety Device for, 375.
——— Lamp, Efficiency of, 170 to 233.
——— Lamp, Half Wire-Guard for, 157.
——— Lamp, Multiple-Series Distribution of, 376, 377.
——— Lamp, Street Fixture for, 378.
——— Electric Lamp, Film Cut-Out for, 376, 377.
——— Electric Lamp, Physics of, 65 to 82.

Incandescent Filament, Brilliancy of, 168.
——— Filaments, Evaporation of, 176, 177.
——— Filament, Total Candle-Power of, 168, 169.
——— Filaments, Surface Activity of, 165.
——— Lamp, Full Wire-Guard for, 158.
——— Head-Lights for Ships, 409.
——— Lamps, Early, 19, 20.
——— Lamps, Miniature, 404, 405.
——— Lamps, Miscellaneous Applications of, 402 to 417.
——— Lamp, Varying Candle-Powers of, 209.
——— Lighting, Early History of, 18 to 42.
——— Lighting, Fire Risks of, 13, 14.
——— Lighting, Alternating-Current Circuit for 386 to 401.
——— Lighting, Decorative Effects in, 416, 417.
——— Side-Lights for Ships, 409.
——— Signal Lights for Ships, 409.
——— Stern-Lights for Ships, 409.
——— Filament, Emissivity of, 80, 81.
——— Filament, Temperature of, 82, 174.
Intake Wires, 280.
Intensity, Luminous, 69.
———, Maximum Horizontal, 207.
Interior Conduit Joints, 263.
——— Conduit Junction Boxes, 264.
——— Conduit, Junction Boxes for, 265.

Interior Conduits, 261, 262.
Isolated Lighting Plants, 324 to 333.
——— Plant, Smashing Point of Lamps, 194, 195.
——— Plants, 324 to 333.
——— Plants, Quadripolar Generator for, 330, 332.

J

Jewellery, Electric, 413.
Joints of Filament with Leading-in-Wires, Bolt and Nut Type, 108.
———, Butt Joint Type, 111.
———, Socket Type, 108.
———, Interior Conduit, 263.
Joule, 60.
Joule-per-Second, 63.
Junction Boxes, 300 to 303.
——— Interior Conduit, 264.

K

Keyless Wall-Socket, 138.
Key-Socket Push Button, 145.
——— Wall-Socket, 139.
Keys for Sockets, 140 to 142.
King, 27.
Kosloff, 29.
Konn, 29.
Konn's Incandescent Lamp, 30, 31.

L

Lamp Adapter, 133.
———, Age-Coating of, 178.
——— Bases, 131, 132.
——— Bracket, 243.
——— Chamber, Blackening of, 178.
———, Filament Shadows on, 179.
———, Machine Sealing of, 121.
———, Sealing Off of, 127, 128.
———, Steam-Tight, 161, 162.
——— Cord, Flexible, 240.
——— Cords, Silk, 248.
——— Distribution, Series-Multiple System of, 221.
———, Distribution Systems of, 209 to 236.
——— Filament, Relation Between Efficiency, Candle-Power, and Surface Activity of, 180.
——— Fittings, 135 to 162.
——— Guard, Portable, 160.
———, Incandescent, 163.
———, Leading-in-Wires of, 100.
———, Mean Spherical Candle-Power of, 207.
——— Pendant, Adjustable, 240.
———, Flexible, 239, 240.
———, Portable Incandescent, 238.
——— Post for Incandescent Lamps, 380, 381.

Lamp Reflector and Shade, Corrugated, 155.
——— Renewals, Rules for Best Commercial Results, 196, 197, 198.
——— Shades, Glass, 156.
——— Shadows, 149 to 156.
———, Smashing Point of, 192.
———, Semi-Incandescent, 37, 38, 39,
——— Socket, Temporary, 147.
——— Socket, Weather-Proof, 148.
——— Sockets, 135 to 140.
———, Stopper, 121, 122, 123.
——— Switch, 140.
——— Switches, 267.
———, Twin-Filament, 409.
Lamps, Battery-Incandescent, 407.
———, Connection in Parallel, 210.
———, Diagram of Multiple Connection of, 212.
———, Distribution of by Four- and Five-Wire Systems, 225.
———, Double-Filament, 409.
———, Incandescent, Use of, for Surgical Exploration, 403.
———, Miniature Incandescent, 404.
———, Safety Incandescent, 407.
———, Series Connected, Diagram of, 211.
———, Series Connections of, 210.
———, Spring Socket for, 146.

Lamps, Tree Distribution of, 230.
Law of Illumination, 203.
Leading-in Wires of Lamp, 100.
Life of Filament, Circumstances Governing, 167.
——— of Incandescent Filament, 167.
——— of Lamp and Efficiency, Relation Between, 183.
Light, Actinic Effect of, 208.
———, Monochromatic, 73.
———, Objective, Significance of, 66.
———, Significance of term, 198.
———, Subjective, Significance of term, 66.
———, Two-Fold Use of Word, 66.
———, Unit of Total Quantity of, 201.
Lighting Plant, Isolated, 325.
———, Series Incandescent, 374 to 385.
Load Diagram of Central Station, 347, 348.
——— Factor, 348.
Lodyguine, 28.
Long Life *vs.* Low Efficiency, 185, 186.
Lumen, 201.
Luminiferous Ether, 67.
Luminous Frequency, 68.
——— Frequency, Effect of Temperature on, 68, 69.
——— Intensity, 69.
——— Intensity, Standards of, 200.

Luminous Intensity, Unit of, 199.
——— Source, Candle-Power of, 199.
Lux, 202.
Lux-Second, 208.

M

Machine Seal of Lamp Chamber, 121.
Main Conductors, Overhead, 284.
——— Cut-Outs, 277.
——— Switch, 277.
——— Tubes, 293.
Mains, 250.
———, Street, 284 to 303.
———, Three-Wire, 276.
———, Two-Wire, 276.
Man Holes, 287.
Marine Switch for Lamps, 412.
Maximum Horizontal Intensity, 207.
Mean Spherical Candle-Power, 207.
Mechanical Air Pump, 125.
Mercury Pump, 125.
——— Pump, Geissler Type, 127.
Mercury Pump, Sprengel Type, 127.
Metallic Half-Shade for Incandescent Lamp, 151.
——— Lamp Shades, 149 to 153.
Meter, Electrolytic, 335 to 339.
Meters, Electric, 334, 344.

Methods for Overcoming Inequality of Feeder Loads, 233, 236.
Mil-Foot, Circular, Definition of, 54.
Miscellaneous Applications of Incandescent Lamps, 402 to 417.
Molecular Transfer of Heat, 76, 77.
Monochromatic Light, 73.
Molding, Dummy, 260.
———, Picture or Ornamental, 259.
———, Section of, 258.
———, Three-Wire, 257.
Mounting of Filament, 99, 100.
Movable Arm for Bracket Lamp, 246.
Multiple and Series Systems of Lamp Distribution, Relative Advantages of, 212 to 220.
——— Connected Lamps, Diagram of, 212.
——— Series System of Distribution of Incandescent Lamps, 376, 377.
Multipolar Generators, 316.

N

Negative Plate of Storage Cell, 351.
——— Pole, 44.
——— Terminal, 44.
Neutral Conductor of Three-Wire System, 222.
Non-Luminous Heat, 10.

O

Occluded Gas Process, 128, 129.
Ohm, 57.
———, Definition of, 50.
Ohm's Law, 57.
Open-Circuit, 46.
Ornamental Moulding, 259.
Outlet Boxes, 266.
Output Wires, 282.
Overhead Main Conductors, 284.
——— Wires, 284.
Overload Switch for Storage Battery Switchboard, 366 to 370.

P

Panel Reflectors, 153, 154.
Parallel Conductors, 248.
Parchmentizing Process, 85, 86.
Pendant Lamp, 244.
Petrie, 24.
Phot, 208.
Physics of the Incandescent Lamp, 65 to 82.
Plant for Isolated Lighting, 325.
Plants, Isolated, 324 to 333.
Plates or Elements of Storage Cell, 351.

Platinum-Wire Incandescent Lamps, Requisites for, 21, 22.
—— Wire, Use of, for Sealing-in Lamp Chamber, 105.
Plug, Cut-Outs, 277.
Point, Distributing, 280.
——, Feeding, 298.
Pole, Negative, 44.
——, Positive, 44.
Portable Electric Incandescent Lamp, 238.
—— Lamp Guard, 160.
Positive Plate of Storage Cell, 351.
—— Pole, 44.
—— Terminal, 44.
Pressure, Electric, 45.
——, Unit of, 46.
—— Switch for Storage Battery Switchboard, 364.
—— Wires, 298.
Primary Circuit of Transformer, 393.
Pulsating Electric Current, 386.
Pump, Mercury, 125.
Push-Button Key Socket, 145.

Q

Quadripolar Generator for Isolated Plants, 330, 332.

R

Radiation of Glow-Worm and Fire-Fly, 78.
——— of Heat, 75.
Rate-of-Doing-Work, 62.
Radiation, Selective, 79.
Rays, Ultra-Violet, 70.
Reactive Coil, 398.
Recording Wattmeter, 341 to 349.
Reflector Shade for Incandescent Lamp, 152.
Regulators, Feeder, 233.
Requisites for Artificial Illuminants, 6.
Residual Atmosphere, Lamp Chambers Intentionally Provided with, 130.
Resistance, Electric, 49.
——— Electric, Unit of, 50.
——— for Feeder Equalizer, 234.
———, Specific, 53.
Resistivity, 53.
——— of Conductors, Effect of Temperature on, 55.
——— of Insulators, Effect of Temperature on, 55.
Reynier, 38.
Reynier's Semi-Incandescent Lamp, 38, 39.
Reynier-Werdermann's Incandescent Lamp, 41, 42.
Risers, 250.

S

Safety Incandescent Lamps, 407.
—— Fuse, 274.
Sawyer, 33.
Sawyer's Incandescent Lamp, 34.
Safety Device, Automatic, for Incandescent Lamp, 375.
Screw Cleats, 256.
Seal, Machine, of Lamp Chamber, 121.
Sealing-in of Filament, 118.
Sealing-off of Lamp Chamber, 127, 128.
Secondary Cell, 352.
—— Circuit of Transformer, 394.
Selective Absorption of Light Radiations, 71, 72.
—— Radiation, 79.
Semi-Incandescent Lamp, 37, 38, 39.
Series and Multiple Distribution, Relative Advantages of, 212 to 220.
—— Arc and Incandescent Lamp Circuit, 383.
—— Connected Lamps, Diagram of, 210.
—— Connection, 47.
—— Connections of Lamps, 210.
—— Incandescent Lighting, 374 to 385.
—— Multiple System of Lamp Distribution, 221.
Service Wires, 250.

INDEX. 443

Shades for Incandescent Lamps, 149 to 156.
Shadows, Filament, 179.
Ship Lighting by Incandescent Lamps, 412.
Ships, Incandescent Head-Lights for, 409.
———, Incandescent Side-Lights for, 409.
———, Incandescent Stern-Lights for, 409.
Short-Circuit, 272.
Short Life *vs.* High Efficiency, 185, 186.
Signal Lights for Ships, Incandescent, 409.
Signs, Illuminated Electric, 414.
Silk-Covered Conductors, 248.
——— Lamp Cords, 248.
Single-Pole Switch, Simple Form of, 268.
——— Pole Switches, 267.
Six-Pole Generator for Central Station, 217 to 320.
Smashing Point of Lamp, 192.
——— Point of Lamp from Central Station Standpoint, 193.
——— Point of Lamp from Consumers' Standpoint, 194.
——— Point of Lamp from Isolated-Plant Standpoint, 194, 195.
Socket Keys, 140 to 142.
Sockets, Simple Form of, 137.
Solid Conductors, 247.
——— Wires, 246.

Sparking at Commutator, 317.
Specific Resistance, 53.
Spotted Filament, 112.
Sprengel Type of Mercury Pump, 127.
Spring Socket for Lamps, 146.
Squirted Filaments, 88, 89, 97.
Starr, 27.
——— King Incandescent Lamp, 27, 28.
Standard Candle, French, 200.
——— of Luminous Intensity, 200.
Storage Cell Tester, Simple Form of, 371, 372.
Stations, Central, 304.
Steadiness of Light, Effect of Frequency on, 388.
Steady Electric Current, 386.
Steam-Tight Lamp Chamber, 161, 162.
Step-Down Transformer, 395.
Step-Up Transformer, 395.
Stern-Lights for Ships, Incandescent, 409.
Stopper Lamp, 121, 122, 123.
Stopper-Mounted Filaments, 122, 123.
Storage-Battery Switchboard, 363 to 369.
——— Batteries, 345 to 373.
——— Capacity, 349.
——— Cell, Efficiency of, 2, 372.
——— Cell, Energy Storage Capacity of, 361.
——— Cell, for Central-Station Work, 359.
——— Cell, Energy Efficiency of, 372, 373.

Storage Cell, Plates or Elements of, 351.
―――― Cell, Negative Plate of, 351.
―――― Cell, Positive Plate of, 351.
―――― Cell, Voltmeter and Electrodes, 370.
Stranded Wires, 246.
―――― Conductors, 247.
Street Incandescent Lamp Fixture, 381, 382.
―――― Fixture, for Series-Incandescent Lamp, 378.
―――― Lamps, Series Connected for Use on Alternating-Current Circuits, 399.
―――― Mains, 284 to 303.
Sub Mains, 250.
Subways, 285, 286.
Sunlight, Analysis of, 74.
―――― Color Values of Artificial Illuminants, 8, 9, 12.
――――, Frequencies Present in, 70.
Supply Conductors, 250.
Surface Activity of Incandescing Filament, 80, 81, 165.
―――― Activity of Lamp Filament, 180.
―――― Activity of Positive Crater of Arc, 166.
Surgery, Use of Incandescent Lamps in, 403.
Suspended Lamps, Wire Guards for, 159.
Switch, Automatic, 273.
――――, Flush, 272.
―――― for Lamps, 140.

Switch, Main, 277.
———, Marine, for Lamps, 412.
———, Pressure, for Storage-Battery Switchboard, 364.
———, Overload, for Storage-Battery Switchboard, 365, 366 to 370.
Switchboard for Central Station, 305 to 308.
Switches, Double-Pole, 267.
——— for Lamps, 267.
———, Single-Pole, 267.
System, Feeder, 253.
———, Three-Wire, of Lamp Distribution, 221 to 224.
Systems of Lamp Distribution, 210.

T

Taps, 251.
Temperature, Effect of, on Luminous Frequencies, 68, 69.
———, Effect of, on Resistivity of Insulators, 55.
——— of Incandescing Filament, 82, 174.
Temporary Lamp Socket, 147.
Terminal, Negative, 44.
———, Positive, 44.
Tester for Storage Cells, Simple Form of, 371, 372.

Thermostat, 339.
Three-Wire Mains, 276.
―――― Moulding, 257.
―――― System, Neutral Conductor of, 222.
―――― System of Feeder Distribution, 231, 232.
―――― System of Distribution, 221 to 224.
Time Illumination, Unit of, 208.
Total Candle-Power of Incandescent Filament, 168, 169.
Transformer, Alternating-Current, 393.
――――, Primary Circuit of, 393.
――――, Step-Up, 395.
――――, Secondary Circuit of, 394.
Transformers, Effect of Size and Weight on Cost and Efficiency of, 397, 398.
―――― Efficiency of, 395.
―――― Step-Down, 395.
Trans-Illumination, 403.
Tree Distribution of Lamps, 230.
Twin Conductors, 248.
―――― Filament Lamp, 409.
Twisted Double Conductor, 248.
Two- and Three-Wire Systems of Lamp Distribution, Relative Economy of, 223, 224.
Two-Wire Mains, 276.
Tube, Underground, 292.
Tubes, Feeder, 298, 299.

U

Ultra-Violet Rays, 70.
Underground Tube, 292.
Unit Generator, 308.
—— of Activity, 63.
—— of Electric Activity, 63.
—— of Electric Flow, 56.
—— of Electric Power, 63.
—— of Electric Pressure, 46.
—— of Electric Quantity, 56.
—— of Electric Resistance, 50.
—— of Illumination, 202.
—— of Illumination, Intensity of, 199.
—— of Time Illumination, 208.
—— of Total Quantity of Light, 201.
—— of Work, 59.
Universal Ether, 67.

V

Violle, 200.
Volt, 46.
—— Ampere, 63.
—— Coulomb, 61.
—— Coulomb-per-Second, 63.
Voltaic Arc, 18.
—— Battery, 47.

Voltmeters, 306.
——— and Electrodes for Storage Cell, 370.

W

Wall Socket, Key, 139.
——— Socket, Keyless, 138.
Watt-Hours, 335.
Wattmeter, Recording, 341 to 349.
Weather-Proof Lamp Socket, 148.
Werdermann, 40.
Werdermann's Semi-Incandescent Lamp, 41.
Wire-Guards for Suspended Lamps, 159.
Wires, In-Take, 280.
———, Overhead, 284.
———, Pressure, 298.
———, Service, 250.
———, Solid, 246.
———, Stranded, 246.
Wiring and Fixtures for Houses, 237 to 283.
——— Cleat, 254.
Wooden Cleats, 255.
Work, 59.
———, Concealed, 260.
———, Unit of, 59.

www.ingramcontent.com/pod-product-compliance
Lightning Source LLC
Chambersburg PA
CBHW032006300426
44117CB00008B/913